P9-DHI-497

By Michael T. Osterholm, PhD, MPH

Living Terrors with John Schwartz

DEADLIEST ENEMY

[illegible]

[illegible]

Unnatural Death

With John Douglas:

Mindhunter

Journey into Darkness

Unabomber

Obsession

The Anatomy of Motive

The Cases That Haunt Us

Law & Disorder

With C. J. Peters, MD:

Virus Hunter

FICTION

Einstein's Brain

Unnatural Causes

Blood Race

The Edge

With John Douglas:

Broken Wings

BY MICHAEL T. OSTERHOLM, PHD, MPH

Living Terrors with John Schwartz

BY MARK OLSHAKER

NONFICTION

The Instant Image

With John Douglas:

Mindhunter

Journey into Darkness

Unabomber

Obsession

The Anatomy of Motive

The Cases That Haunt Us

Law & Disorder

With C. J. Peters, MD:

Virus Hunter

FICTION

Einstein's Brain

Unnatural Causes

Blood Race

The Edge

With John Douglas:

Broken Wings

DEADLIEST ENEMY

OUR WAR AGAINST KILLER GERMS

Michael T. Osterholm, PhD, MPH
and Mark Olshaker

LITTLE, BROWN AND COMPANY
New York • Boston • London

Little, Brown and Company
Hachette Book Group
1290 Avenue of the Americas, New York, NY 10104
littlebrown.com

First Edition: March 2017

ISBN 978-0-316-34369-5

LCCN 2016957620

10 9 8 7 6 5 4 3 2 1

To the three people who uniquely have influenced my path in life with faith and love. In their own ways, each has taught me to learn from my yesterdays and today and to dream about a better tomorrow:

The late Laverne Keettel Hull, who gave me the road map of life when I was a young boy;

David "Doc" Roslien, who has inspired me for more than forty-five years to dream, using the confluence of science and policy as my North Star;

Dr. Kristine Moore, without whose professional support and counsel I would not be where I am today.

MICHAEL OSTERHOLM

To my brother, Dr. Jonathan S. Olshaker, who has devoted his life to the front lines of everything we're fighting for, with love and admiration.

MARK OLSHAKER

Contents

Introduction 3

Chapter 1: Black Swans and Red Alerts 7

Chapter 2: Annals of Public Health 21

Chapter 3: White Coats and Worn Shoes 30

Chapter 4: The Threat Matrix 49

Chapter 5: The Natural History of Germs 57

Chapter 6: The New World Order 64

Chapter 7: Means of Transmission: Bats, Bugs, Lungs, and Penises 72

Chapter 8: Vaccines: The Sharpest Arrow in Our Quiver 80

Chapter 9: Malaria, AIDS, and TB: Lest We Forget 98

Chapter 10: Gain of Function and Dual Use: The Frankenstein Scenario 111

Chapter 11: Bioterror: Opening Pandora's Box 125

Chapter 12: Ebola: Out of Africa 144

Chapter 13: SARS and MERS: Harbingers of Things to Come 159

Chapter 14: Mosquitoes: Public Health Enemy Number One 178

Chapter 15: Zika: Expecting the Unexpected 205

Chapter 16: Antimicrobials: The Tragedy of the Commons 215

Chapter 17: Fighting the Resistance 237

Chapter 18: Influenza: The King of Infectious Diseases 254

Chapter 19: Pandemic: From Unspeakable to Inevitable 268

Chapter 20: Taking Influenza off the Table 286

Chapter 21: Battle Plan for Survival 299

Acknowledgments 321

Index 327

Humanity has but three great enemies: fever, famine and war; of these by far the greatest, by far the most terrible, is fever.

— Sir William Osler, MD

A good hockey player plays where the puck is. A great hockey player plays where the puck is going to be.

— Attributed to Wayne Gretzky

DEADLIEST ENEMY

Introduction

When I was the state epidemiologist of Minnesota, a few people in the media started calling me "Bad News Mike" because often when public officials or corporate leaders got a call from me, I was unlikely to tell them anything they wanted to hear. In a story with that title by Kermit Pattison in *Mpls St Paul* magazine, the subhead read: "Headstrong and outspoken, the state's epidemiologist insists he's only a messenger from the germ front. Whatever he is, the message isn't good."

Well, I don't know about the "headstrong" charge, but I certainly have to plead guilty to being "outspoken." That's because I believe in what I call *consequential epidemiology*. That is, by attempting to change what could happen if we don't act, we can positively alter the course of history, rather than merely record and explain it retrospectively. Because of the accomplishments in the 1960s and 1970s of two of the giants of public health, Drs. Bill Foege and the late D. A. Henderson, aided by literally thousands of others, countless millions yet unborn will be spared the devastation of smallpox. Opportunities for such life-altering good are still out there, if we only recognize them and have the collective will to act.

This book results from my participation, observations, concerns, outbreak investigations, studies, programs, and policy development on the front lines of the major public health issues of our time. They involve toxic shock syndrome, AIDS, SARS, antibiotic resistance, foodborne diseases, vaccine-preventable diseases, bioterrorism, zoonotic diseases (those transmitted from or to animals and humans) including Ebola, and vector-borne diseases (those transmitted by mosquitoes, ticks, and flies, such as dengue and Zika viruses). Each experience or encounter—local, regional, national, or international—has informed and shaped my thinking, each has taught me a critical lesson about how we deal with our deadliest enemy, and each has focused the lens through which I approach public health.

Because, in fact, infectious disease is the deadliest enemy faced by all of humankind. True, infection is far from the only type of illness that affects each of us, but it is the only type that affects us collectively, and sometimes on a mass scale. Heart disease, cancer, even Alzheimer's, can have devastating individual effects, and research leading to cures is laudable. But these diseases don't really have the potential to alter the day-to-day functioning of society, halt travel, trade, and industry, or foster political instability.

If there is any particular theme to my career, it has been connecting disparate dots of information and making them into a coherent line to the future. For example, I both wrote and lectured as early as 2014 that the appearance of the Zika virus in the Americas was just a matter of time. Before a doubting professional audience at the National Academy of Medicine in 2015, I predicted that MERS would soon appear in a major city outside the Middle East. (It did, in Seoul, South Korea, just months later.)

I don't claim any unique skills. Foreseeing issues and potential threats should be a matter of standard practice in public health.

When I established CIDRAP, the Center for Infectious Disease Research and Policy that I now head at the University of Minnesota, I was mindful of the fact that without policy, research has nowhere to go. Another way of saying this is that we tend to go from crisis to crisis without ever anticipating them or finishing the job in the end.

Science and policy must intersect to be effective. Therefore, throughout this book, we will seldom talk about realized or needed advances in the science of disease prevention without always giving equal consideration to what to *do* with those advances.

What we aim to give you here is a new paradigm for considering the threats posed by infectious disease outbreaks in the twenty-first century. While we will deal with the broad range of communicable illness, we will concentrate on identifying and exploring those maladies with the potential to disrupt the social, political, economic, emotional, or existential well-being of large regions, or even the entire planet. And while morbidity and mortality are certainly prime considerations, they are not the only ones. The current reality is that a few confirmed cases of smallpox anywhere in the world would create more sheer panic than do many thousands of malaria deaths in Africa alone.

That is, we don't always make rational distinctions between what is likely to kill us and what is likely to hurt us, scare us, or simply make us uncomfortable. As a result, we don't always make rational decisions about where to put our resources, where to direct our policy, and, frankly, where to direct our fear. As we write this, much of the Western world is greatly worried by the spread of Zika virus and its association with microcephaly, other birth defects, and Guillain-Barré syndrome. Yet over the past few years, dengue virus, which is spread by the very same mosquito, has killed far more people in the same region with hardly a blip on the public radar. Why? Probably because there are few situations as dramatic and horrifying as babies being born with small heads and facing uncertain lives of disability. It is every parent's worst nightmare.

We will be invoking two metaphors for disease throughout this book. One is crime and the other is war, and both are apt because in various ways, our struggle against infectious disease resembles both of these horrors. In the investigation and diagnosis of outbreaks, we are like detectives. In our response, we must be like military strategists. Just as we will never eliminate either crime or war, we will never eliminate disease. And just as we engage in an ongoing war against crime, we are constantly battling disease.

In the first six chapters, we will present the stories, cases, and backdrop that will provide context for the rest of the book. From that point on, we will discuss what I consider our most pressing threats and challenges as well as practical means to take them on.

In 2005, I wrote an article for the journal *Foreign Affairs* entitled "Preparing for the Next Pandemic." I concluded with the following warning:

> This is a critical point in history. Time is running out to prepare for the next pandemic. We must act now with decisiveness and purpose. Someday, after the next pandemic has come and gone, a commission much like the 9/11 Commission will be charged with determining how well government, business, and public health leaders prepared the world for the catastrophe when they had clear warning. What will be the verdict?

In the eleven years that have passed since I wrote those words, I don't see that much has changed.

We could try to scare you out of your wits with bleeding eyeballs and inner organs turned to mush as some books and films have attempted to do, but in the vast majority of instances, those images are a misrepresentation and not relevant. The truth and the reality should prove sufficiently concerning to scare us all *into our wits.*

I'm not trying to give either an optimistic or a pessimistic spin on the challenges in facing our deadliest enemy. I'm trying to be realistic. The only way we are going to confront and deal with the ever-present threat of infectious disease is to *understand* those challenges so that the *unthinkable* does not become the *inevitable.*

CHAPTER 1

Black Swans and Red Alerts

There's something happening here.
What it is ain't exactly clear.
— BUFFALO SPRINGFIELD

Who? What? When? Where? Why? How?

Just like reporters and police detectives, this is what public health epidemiologists—disease detectives—always want to know: as many pieces of the "How did this happen?" puzzle as possible; the components that help us tell the story. That's what epidemiology—in fact, all of diagnostic medicine—is about: connecting the dots and putting together a coherent story. And only then, once we sufficiently know and understand the story, can we begin to confront the problem or challenge. As medical detectives we can sometimes stop a disease outbreak cold without understanding all the pieces of a complex puzzle, like finding that a certain food item is making people sick even though we don't know how that food got contaminated. But the more we can find out, the better equipped we are to solve the mystery and make certain that similar disease problems don't happen in the future.

On a day I will never forget, there were about ten of us sitting around the table in the Director's Conference Room at the Center for Disease Control in Atlanta—later renamed the Centers for Disease Control and then again the Centers for Disease Control and Prevention. None of us knew what to make of the cases that had just been presented to us as we went through the mental checklist.

The *what:* in one cluster, *Pneumocystis carinii* pneumonia (PCP)—a rare parasitic infection that causes a life-threatening pneumonia and usually occurs only in people with compromised immune systems. And in the other, Kaposi's sarcoma (KS)—a disfiguring malignant tumor now known to be caused by human herpesvirus-8 (HHV-8) and also more frequent in people with immune system problems. It starts as little red and bluish black lesions on the skin or in the lining of the mouth, nose, or throat. The lesions grow into very painful raised tumors and often spread to the lungs, digestive tract, and lymph nodes.

The *when:* right as we sat there—June 1981.

The *where:* The PCP cases were being found primarily in the Los Angeles area and the KS cases in the New York City area.

The *who:* two clusters of young, otherwise healthy gay men on opposite sides of the country.

The *why* and *how:* Those were the mysteries.

Because, we all knew, *these rare, arcane diseases shouldn't be happening in this patient population.*

Dr. James Curran sat at the head of the table in the long, narrow room paneled in dark wood. He was with what was then called the STD Division—sexually transmitted diseases—and his team was working with the CDC's Viral Hepatitis Branch in Phoenix. I was interested in hepatitis B and was doing studies on how healthcare workers at a single hospital in Minneapolis had become infected. More than eighty such cases had occurred in a fourteen-month period, including that of a young physician who had died as a result of his work-acquired hepatitis infection.

Jim is one of the brightest guys in our business and someone never afraid to speak his mind. I had once considered taking a job in his division at the CDC. Now he was setting up a study on a new, not yet approved hepatitis B vaccine in gay men in several cities across the United States. Gay men were at high risk, due to the significant possibility of transmitting the virus through anal sex, a risk heightened for those with multiple partners.

Dr. Bill Darrow, an STD Division expert on the behavioral aspects of infectious disease, and Dr. Mary Guinan, MD, PhD, a leading virus expert with the STD Division, were also at the meeting.

Dr. Dennis Juranek of the Division of Parasitic Disease was there and had been quite involved with the early information gathering for these cases. Since PCP was so rare in the United States, the manufacturer of the chief drug used to treat it worldwide, pentamidine, had not wanted to go through the time and expense of the full Food and Drug Administration approval process. Therefore, the CDC was the only place in the United States that could stock it, as an investigational, unlicensed drug. Dr. Wayne Shandera, who helped monitor disease outbreaks from Los Angeles as part of the Epidemic Intelligence Service (EIS), was on the speakerphone. EIS is the CDC's training program for new epidemiologists and other public health professionals, who are sent around the nation and the world to investigate mysterious and potentially threatening disease outbreaks.

For a twenty-eight-year-old epidemiologist from the Midwest, working with such distinguished and dedicated people and being there at the CDC was like beaming up to the mother ship. I was grateful that Jim had invited me to this meeting, even as a small-bit player. As chief of the Acute Disease Epidemiology Section of the Minnesota Department of Health, I was actually at the CDC for another reason—a meeting on toxic shock syndrome (TSS), a condition I had been actively investigating for almost a year. Because of that, my experience with public health disease surveillance related to unexplained outbreaks, and the fact that I happened to be in the building, Jim invited me to help provide a perspective from the field. In addition, I had led our team at the Minnesota Department of Health in investigating several recent large outbreaks of another type of viral hepatitis in gay men. That illness is now known as hepatitis A.

It was against this public health backdrop and recent investigative experience that I faced the current mystery with the others in the CDC Director's Conference Room.

Details had been published, employing the dispassionate language of science, in the June 5, 1981, issue of *MMWR*—the *Morbidity and Mortality Weekly Report*—the CDC's dispatch of diseases important to the public:

> In the period October 1980–May 1981, 5 young men, all active homosexuals, were treated for biopsy-confirmed *Pneumocystis carinii* pneumonia at 3 different hospitals in Los Angeles, California. Two of the patients died. All 5 patients had laboratory-confirmed previous or current cytomegalovirus (CMV) infection and candidal mucosal infection. Case reports of these patients follow.

The report described five men, ages twenty-nine to thirty-six, four of whom were previously healthy and the fifth having been successfully treated for Hodgkin's lymphoma three years earlier. CMV is a common virus that most carriers don't know they have, because it generally doesn't cause any symptoms. Since it spreads from person to person via bodily fluids—saliva, blood, urine, and semen—and because people share more fluids when they have multiple partners, and also because anal intercourse is much more likely to cause small abrasions and resultant bleeding than vaginal intercourse, it was often noted in sexually active gay men. The term of art in those days was MSM—men who have sex with men. But CMV was known to cause various health problems in individuals with compromised immune systems. The candida infection these men presented with could indicate some sort of immunosuppression. Patient 4, the youngest of the cohort and the one who had had Hodgkin's disease, was one of the two who had died. He had been treated with radiation. Had that suppressed his immune system? Had the cancer itself had some effect? What about the other four?

Particularly confounding was that these two conditions—*Pneumocystis carinii* pneumonia in LA and Kaposi's sarcoma in New York—were not "perpetrators" any medical detective would expect to discover at such a "crime scene." PCP was caused by a parasite that, in general, is easily neutralized by the human immune system. KS in this part of the world tends to show up in old, otherwise frail and sickly men.

As *MMWR* soberly noted:

> *Pneumocystis* pneumonia in the United States is almost exclusively limited to severely immunosuppressed patients. The occurrence of pneumocystis in these 5 previously healthy individuals without a clinically apparent underlying immunodeficiency is unusual.

So why were we seeing these two medical anomalies in groups of healthy young men on both coasts? What were the known causes of immunosuppression?

We went through the list of usual and unusual suspects—what physicians refer to as the differential diagnosis.

There was some speculation that it could be related to Epstein-Barr virus (EBV), generally transmitted through oral and genital secretions and bodily fluids. Often, EBV causes no symptoms at all, but it is one of the prime causes of infectious mononucleosis, which when I was in school was known informally as the "kissing disease." EBV is also associated with more serious conditions, including Hodgkin's and Burkitt's lymphomas and a variety of autoimmune diseases. Some scientists have speculated that it triggers chronic fatigue syndrome, though the association has never been proven.

Theories were running rampant—everything from the idea that none of these cases were related to the appearance of a new, highly infectious disease.

"Most of us thought it was a sexually transmitted agent, but we didn't know what," Jim Curran recalled.

Could there be some blood-borne microbe that was promoting these conditions? Maybe there was a chemical these men had intentionally or inadvertently ingested. We thought it sounded like an infectious disease, but at that point, we couldn't be sure.

There was a significant cohort of the gay community in a number of major cities, New York and LA included, that was sexually active with numerous partners, often on the same day. So one of the favored methods for achieving and maintaining an erection, and enhancing sexual sensation, was through sniffing amyl nitrite "poppers." Were the chemicals lingering in the system and causing these weird effects? It didn't seem likely, but we weren't ruling anything out.

And the big question: Were these two clusters related, or was the commonality of sexually active gay men merely a fluke? Most people have heard the old diagnostic aphorism, *Common things occur commonly. Uncommon things do not. When you hear hoofbeats, think of horses before you think of zebras.* So was this a zebra, or simply two unrelated horses?

The first critical step would be what we call "case surveillance," and it is just as important as a police detective's surveillance of a possible suspect. Because of my own recent experience with toxic shock syndrome, the group assembled in that conference room asked me how I thought they could enhance surveillance in New York and LA and where else they should look for similar cases. Did it make sense to concentrate on clinics that handled a lot of sexually transmitted diseases? What about pulmonologists' offices for possible cases of PCP and dermatologists' for KS?

Those ideas made sense, but I thought we would likely get the most information quickly by conducting a survey among doctors in the areas of LA and New York City with large populations of gay men to see if any of them were seeing cases like these. Even if these cases were caused by a single infectious microbe or ingestion of a chemical that undercut the immune system and occurred in other cities and among heterosexuals, the "hot spots" for finding more cases seemed to be among gay men in LA and New York City.

I walked out of the meeting wondering if there was really anything to worry about or if these cases were just the kinds of random incidents that happen in our business. Would one or both of these small clusters turn out to be medical anomalies that quickly faded from view? Would they be mysteries with neat explanations? That was certainly what Jim was hoping for; as he said, "Identify. Treat. Over."

Or were we seeing a genuine black swan event, one that would become an all-hands-on-deck red alert?

The term "black swan" was introduced by Nassim Nicholas Taleb, author and scholar, to explain certain rare occurrences in financial markets. In his 2007 book, *The Black Swan,* he extended the concept to explain unusual high- or extreme-impact and difficult-to-predict events in the larger world.

None of us around the table that day in Atlanta realized that we were bearing witness to an epochal moment in history: the world's transition into the era of AIDS. Jim Curran would remain the CDC's point man on the disease, and it would transform his career.

Jim subsequently set up a CDC task force to explore this new condition, tentatively labeled Kaposi's Sarcoma and Opportunistic Infections. At about the same time as the establishment of the task force and the publication of the first *MMWR* report, the CDC began receiving an unprecedented number of requests from physicians for pentamidine to treat young men afflicted with PCP, especially in New York. Even though no one knew what was causing the condition, Jim and his colleagues knew it was time for the CDC to develop a case definition.

The case definition is critical in identifying a disease and trying to figure out what to do about it. Once a disease has been described in this way, the CDC's own investigators, state and local health department officials, hospital emergency room personnel, and all other physicians and healthcare workers can begin ruling in and ruling out individuals they see.

"The cases were so unusual," Jim recalled, "that we had to have a specific definition. Then we focused on very specific active surveillance, so we were able to say, 'This really is increasing. It's focal, but it's spreading.' "

As soon as the media picked up the story of these strange new disease outbreaks, the CDC was flooded with calls describing similar symptoms. By the end of 1981, 270 cases of severe immunodeficiency had been reported in gay men. Of those, 212 had died. In the first year or so of surveillance, the condition was seen mostly in gay men and intravenous drug users.

The next year, the disease estimate was in the tens of thousands. Jim says, "The problem was that the first few years, we were always underestimating but being accused of overestimating."

It was when symptoms started showing up in people who didn't fit the profile that the investigation turned a critical corner. Jim recalls, "We started seeing transfusion recipients with *Pneumocystis* pneumonia, and we were pretty convinced they weren't gay and had no other risk factors. We saw it in children with hemophilia. Then we were able to convince ourselves and others of the logic of who got it and who didn't. And that was really important. When we saw three hemophilia cases in one week, we knew the agent had to be in the blood supply, and it had to be a yet-unrecognized virus."

In September 1982, under Jim's leadership, the CDC first used the term "acquired immune deficiency syndrome," which was defined as "a disease at least moderately predictive of a defect in cell-mediated immunity, occurring in a person with no known case for diminished resistance to that disease." Jim had pushed for the adoption of the AIDS acronym because he thought it was critical to have a name that was easy to remember and would have the same label throughout the world.

The next month, *MMWR* published its first guidelines on AIDS prevention, treatment of patients, and handling of specimens.

AIDS turned out to have all the elements of the greatest public health challenges: on-the-scene medical drama, in-the-lab discoveries, and huge financial, social, religious, ethical, political, and even military impact.

By 1983, lab scientists in the United States and France had determined that AIDS was caused by a retrovirus. On April 23, 1984, Health and Human Services secretary Margaret Heckler held a press conference to say that Dr. Robert Gallo and his colleagues at the National Cancer Institute of the National Institutes of Health had found the cause of AIDS: the retrovirus HTLV-III.

This would be followed in June by Gallo and Pasteur Institute professor Luc Montagnier's joint press conference confirming that the French lymphadenopathy associated virus (LAV) and the American HTLV-III were almost certainly identical and the likely cause of AIDS. It then took until 1986 for the International Committee on Taxonomy of Viruses to officially label the cause of AIDS as human immunodeficiency virus, or HIV.

HIV most likely began in the jungles of Africa as an infection in primates such as monkeys or chimps, and it lingered there for many decades before crossing over into the human population.

As human populations grew in the jungles of Africa, the practice of hunting primates became more common and bushmeat a regular source of nourishment. The virus probably jumped species as people killed, butchered, and had extensive blood contact with infected primates. From there, human-to-human sexual transmission was probably the main means of spread, eventually making it out of the small, isolated groups in the jungle.

This is an instructive model for the proliferation of other infectious diseases as population growth and "progress" create better roads and more mobility while reducing jungle and forestland. As a result, microbes that may have stayed in their particular niches for centuries or longer are now emerging into far larger problems.

But to go back to the April 23 press conference, Margaret Heckler also announced development of a diagnostic blood test and expressed hope that an AIDS vaccine would be ready within two years.

The idea that an AIDS vaccine would be ready that soon struck me as wildly unrealistic. I couldn't fathom where she had come up with that estimate. Two years is a very short amount of time to develop any vaccine, and for the retrovirus that caused AIDS, the time frame seemed virtually impossible.

Once in the cell, the retrovirus hangs around indefinitely. HIV is present in the body fluids of infected individuals, and when the virus enters a person in the form of infected immune cells, for example, in ejaculate, it makes it virtually impossible for antibodies produced by a vaccine or other parts of the normal human immune response to win the earliest battle against the invading virus. With other viruses, vaccines trigger the immune system to identify the invaders and kill them. But the fact that this virus could escape the body's own defenders challenged all notions of how vaccines work.

"There was definitely some premature optimism with mention of the vaccine," Jim comments. "The honest question would be, not *when* there would be a vaccine, but *if* there would be a vaccine."

This didn't mean treatments couldn't be developed that would greatly handicap the virus once it was in the body. In fact, progress on the cocktail of drugs now used to control the disease has been truly remarkable and inspiring. But the key word here is *control,* just as we do with diabetes and other chronic diseases, not *prevent* or *cure.*

In the mid-1980s, while some in the public health community were laser focused on vaccine research, I kept saying in every forum I participated in that we couldn't afford to wait for a vaccine to stop transmission. Preventive measures were essential.

I had a personal stake in this. In 1983, before the American blood supply was routinely screened for HIV, my beloved sixty-six-year-old aunt, Romana Marie Ryan—a nun and teacher in San Francisco—broke her hip when she fell while taking her kindergarten class on a field trip. Her parish priest, Father Thomas F. Regan, often said that she was "magically gifted" in teaching young children.

Aunt Romana had come home to Iowa for a visit in August of 1984. We had a small family reunion at the motherhouse convent in Dubuque. I remember so clearly driving from Minneapolis to Dubuque for a wonderful Sunday afternoon get-together.

It was a beautiful day on the bluffs overlooking the Mississippi River. Sister was her usual joyful, fun, and loving self, the kind of person you cherish being with. But she had been sick lately and her doctors had not been able to pinpoint the cause. I remember she was wearing a long light-green skirt that day; she had given up nuns' garb years earlier. When she was sitting on a patio chair, I noticed she had these terrible-looking red and purple lesions on her lower legs.

Even with my familiarity with KS, I did not put two and two together. She wasn't a gay man and I didn't realize she'd had a blood transfusion during her 1983 hip surgery to fix the broken bone; the doctors had assumed she'd have substantial blood loss, so she had been started on a transfusion at the beginning of the operation. The blood she had received had been contaminated with HIV. And it turned out the transfusion had been unnecessary, as she didn't have extensive bleeding.

Not long after her return to San Francisco, Romana was diagnosed with AIDS. She died of *Pneumocystis carinii* pneumonia in February 1985, spending her last months in agonizing pain. But she never complained, instead praying daily for the HIV-infected man who had donated the blood she had received, and all the others who shared her condition. "I know how they are suffering," Father Regan quoted her as saying. "I am offering what is happening to me so that the doctors will find a cure for this disease."

The virus consumed her body but left her holy and kind soul undiminished. Romana was the closest person to me to date who would die from AIDS. But over the next thirty years this microbial monster would take a number of dear friends and colleagues.

Just days after Secretary Heckler's infamous 1984 press conference, I gave a talk to a Twin Cities gay business group. There were more than two hundred in the audience, many of whom were in denial, believing that in my public pronouncements I had been exaggerating this whole AIDS issue.

During my introduction, the MC stated with excitement and a sense of relief that with Secretary Heckler's announcement of a soon-to-be-available vaccine, this new gay health crisis would soon pass. It was almost as if he were saying I really didn't need to be there after all.

I started my talk with one simple message: I put no credence in Secretary Heckler's statement and did not believe we would see an effective AIDS vaccine in my professional lifetime unless a new technology akin to a "Beam me up, Scotty" machine was discovered. There were a number of boos and shouts from the audience. A few people even got up and walked out. I knew what I said was completely based on the science of retrovirology and epidemiology. But that fact brought no comfort as I stood before this group, one I knew would experience a large number of painful deaths in the months and years ahead if its members didn't heed the message of safer sex and personal prevention. It was one of my classic "Bad News Mike" moments, but the evidence pointed in only one direction.

In 1985, the state of Minnesota became the first government body in the world to make HIV infection a reportable public health condition. We, and several other state and local health departments, had made AIDS—the full-blown disease—reportable the previous year. I led that effort as part of a comprehensive public health program to address HIV infection, just as we would and should for any serious infectious disease threat. HIV-infected persons were assured that their health status would remain confidential and not become public information or shared with their employers as a result of the mandatory reporting. But it was a very unpopular action among most in the gay community.

In 2006, the CDC recommended universal HIV screening, something I had advocated publicly in the mid-1980s—another move on my part that wasn't exactly popular. It wasn't until 2015 that major healthcare providers around the country, including in my own state of Minnesota, came out for universal screening of everyone between the ages of eighteen and sixty-four.

Twenty years after that first mention in *MMWR,* the CDC announced that nearly half a million people had died from AIDS in the United States alone. Yet officials were still writing, "The development of an HIV vaccine is important to control the global epidemic." As of this writing, we still don't have such a vaccine, despite continual promises and expressions of hope from public health officials and laboratory researchers. And it is not for lack of trying.

In 2014, an estimated 36.9 million people around the world

were living with HIV infection, most in sub-Saharan Africa. There are an estimated 2 million new cases a year and 1.2 million deaths. Today, during an average week, there are 30,000 new HIV infections, and 20,000 will die from AIDS in sub-Saharan Africa. As long as new cases exceed the number of deaths, the overall number of people living with HIV infection increases.

The good news is that approximately 15 million HIV cases are currently receiving antiretroviral therapy. The bad news, that almost 22 million cases around the world are not; that's almost 60 percent of the total HIV-infected population. With those 2 million new cases annually, it's fair to say that globally, we no longer have an "AIDS epidemic." HIV infection is still a public health crisis, particularly in sub-Saharan Africa, but now it's what we call "hyperendemic": a really bad public health problem that doesn't go away.

AIDS can serve as a dire warning about the *possible:* a black swan of an infectious disease that seemingly came out of nowhere, unleashing unimagined suffering on an unsuspecting world. As such, it is a classic example of the ongoing tension between horses and zebras, a tension that has defined my professional career and has had a permanent impact on my approach as an epidemiologist.

AIDS is a horror story that haunts all of us in the business. Once we understood what we were dealing with and how it was transmitted, we were unable to stop or warn off much of the behavior and habits that led to its spread. Evidence, knowledge, and logic aren't always enough.

CHAPTER 2

Annals of Public Health

The first step in the evolution of ethics is a sense of solidarity with other human beings.

— Albert Schweitzer, MD

I grew up in Waukon, Iowa, a small farming town in the northeastern corner of the state, home to the venerable Allamakee County Fair and about fifteen miles west of a bend in the Mississippi River. I was the oldest of six children—three boys and three girls—with a physically abusive, alcoholic father. I came home late the night of my senior high school homecoming to find that my father had beaten up my mother and smashed a beer bottle on her head. This was the most severe violence I'd ever seen from him, including that regularly inflicted on my mother, my siblings, and me. It was the only time in my life I physically confronted someone. In fact, though I'm not particularly proud of this, I damn near killed him.

I often quote Sir Winston Churchill's directive "Play for more than you can afford to lose and you learn how to play the game." That night, I played for more than I could afford to lose because I knew at that point that he could never come back into our home again.

Of course, this family crisis was all hush-hush back then, but my father never returned home.

If nothing else, this incident taught me the lifelong lesson of when it's critically necessary to stand your ground, and when it isn't.

Some of my friends have suggested that this background accounts for a need to protect everyone around me. I'm not sure about that. What I do know is that it was in junior high school that I set my life's course.

I had always been interested in science, but I also loved mysteries and read Sherlock Holmes stories voraciously.

My father was a photographer for the local newspapers, the *Waukon Democrat* and *Waukon Republican-Standard,* which were owned by two brothers. The wife of one of them, Laverne Hull, subscribed to *The New Yorker* and gave me copies after she read them. I'm certain she was the only subscriber in Waukon, if not all of northeast Iowa. I was fascinated by a feature called "The Annals of Medicine," written by the wonderfully gifted Berton Roueché. Every time one of his articles appeared, I plunged into the medical mystery he described and imagined myself part of the team of science detectives who solved it. I didn't even know the term "epidemiologist" in those days, but I knew I wanted to be one.

Particularly gratifying to me, in 1988, toward the end of his career, Roueché wrote an "Annals of Medicine" on a thyrotoxicosis outbreak in southwest Minnesota and South Dakota whose investigation I led. It was one of the greatest gifts of my professional life to be able to come full circle with Mr. Roueché.

What is it we do and why do we do it?

Epidemiology is the study of disease in populations, with the aim of preventing disease in people and animals. Public health has an overlapping definition; it refers to those actions taken to achieve the goal of improving health in a given community, whether that community is a small town in Minnesota, the continent of Africa, or the entire planet.

My hero and friend William "Bill" Foege, former director of the CDC, former executive director of the Carter Center, and now a senior fellow and consultant to the Bill & Melinda Gates Foundation, says, "The purpose of public health is to promote social justice," going on to explain, "Its philosophical base is social justice, and its scientific base is epidemiology."

To better explain what he meant, Bill cited Primo Levi, the revered Italian chemist, philosopher, and author whose searing memoir, *Survival in Auschwitz,* is one of the essential Holocaust narratives. Levi said, "When you know how to relieve torment and don't, then you become the tormentor." I have never heard our collective mission more exquisitely stated.

Bill is one of the towering figures in public health—at six feet seven, both literally and metaphorically. Perhaps his greatest achievement was his participation in the global effort to eradicate smallpox, both on the ground and through devising and implementing the "ring strategy" of vaccination—officially known as "surveillance and containment." It is not surprising, then, that when Microsoft founder Bill Gates and his wife, Melinda, decided to dedicate much of their multibillion-dollar fortune to a foundation devoted to world health, they chose Bill Foege as one of their chief advisers. In establishing their foundation, they were pursuing their belief that every child is entitled to a healthy life, to the extent that other human beings can provide it. "It is our responsibility to bring people around the world as close to a level of health as possible," Gates commented.

As a professor in a school of public health, I am often asked by my students how we can prepare to face the overwhelming challenges that epidemic and pandemic diseases pose. My answer is to take a page from Bill Foege's playbook.

Bill cites three tenets to his personal philosophy as it applies to public health, which we would all do well to follow:

First, as confusing and bewildering as things may seem, we live in a cause-and-effect world. So somewhere, the answers are out there.

Second, know the truth—and the first step to knowing the

truth is *wanting* to know the truth, rather than any alternative that seems more satisfying or closer to your own worldview.

Third, not one of us does anything worthwhile on our own.

To these principles, I would add one more: We're all in this together, whether we like it or not. As the great and prescient microbiologist and Nobel laureate Dr. Joshua Lederberg warned us, "The microbe that felled one child in a distant continent yesterday can reach yours today and seed a global pandemic tomorrow." Josh was one of the most influential people in my career until his death in 2008. As a mentor he taught me that one dot is just that: an isolated person, bacterium, virus, parasite, place, or time. But a bunch of dots begins to make a line if they become organized by chance or by design. That's our job in public health: to see the dots before they become a line and do whatever we must so that line never materializes.

One of Bill Foege's lifelong goals was to read all the works of American historians Will and Ariel Durant, especially their epic eleven-volume *The Story of Civilization*. In a conversation at the Rollins School of Public Health at Emory University in Atlanta, he told us how, after the Japanese attack on Pearl Harbor on December 7, 1941, the entire country and much of the world seemed to come together overnight. Since then, he wondered, has there been anything that could trigger a similar coalition of the righteous and committed? The September 11, 2001, terrorist attacks did that initially, many would argue. But the reaction didn't last long, muddled and dissipated as it was by military action that arguably had nothing to do with the attack or threat.

An alien invasion, though, that threatened the entire planet and forced human beings to set aside their differences would do it, the Durants believed.

"Infectious diseases turn out to be a surrogate for an alien invasion," Bill declared. "It's why we were able to do smallpox eradication in the midst of the Cold War. Both sides could see this was an important thing to do."

To take the alien invasion analogy one step further, we would first have to convince the public that extraterrestrials had, in fact, landed on earth. Look at climate change: The science is well established and yet a large percentage of the population refuses to believe it.

The same holds true for infectious diseases. Our task is to convince world leaders, corporate heads, philanthropic organizations, and members of the media that the threat of pandemics and regional epidemics is real and will only continue to grow. Ignoring these threats until they blow up in our faces is not a strategy.

So what *is* the public health agenda?

It is not to prevent death; let's get that one out of the way here and now. That's still impossible. The overall rate of death to birth so far always has been—and as far into the future as we can see, always will be—constant at 100 percent: one death for every birth. The agenda is not even to prevent the so-called leading causes of death. If you could do that, there would still be a top ten causes of death, and I'm certain some of them wouldn't be any better than the ones we have now. What we in the public health sphere are always trying to do is *replace* bad deaths with good deaths; to prevent early and needless death and disease. As medical science and public health capabilities progress, we should continually redefine the unacceptable.

Just about all deaths are sad, and many are tragic. But from a public health standpoint, there are more profound and meaningful differences. A ninety-year-old man with limited mental and physical impairment expiring in his sleep is a good death. A six-year-old child, whether living in the United States or in a country in Africa or Asia, dying of a diarrheal disease is a bad death. The first is a peaceful end to a long and eventful life. The second is the loss of many decades of life and potential and an absence of future generations.

As epidemiologists, we have two goals. The first is to prevent. When that is not possible, the second is to minimize disease and extended disability. Toward that end, we deploy an arsenal of medical countermeasures.

We have several important weapons for prevention: sanitation, including safe water and food and the safe removal of human and animal feces and urine; vaccination; and anti-infectives, which can minimize disease, disability, and, potentially, infectiousness. Vector control is critical for reducing disease-transmitting mosquitoes, ticks, and flies. Then there are ancillary measures, such as disinfecting agents, and infection control in hospitals, nursing homes, and daycare facilities. And there are nonmedical actions, too, including education, attempts to get the public to change certain behaviors, public communications, and quarantine. Guidance on sexual habits and precautions for multiple-partner activity are examples. So is changing burial practices for Ebola fatalities, as we learned during the 2014 outbreak in West Africa.

But the fundamental tool of epidemiology has always been, since long before we had a scientific method for identifying microbes or a germ theory of disease—and, I expect, will always be—observation.

In rural England, by the eighteenth century it had been observed and noted that milkmaids seemed generally to be immune from the scourge of smallpox, which had a mortality rate of at least 30 percent, and often significantly higher. Dr. Edward Jenner speculated that exposure to the similar but far less serious cowpox somehow protected them. In May of 1796, in a now legendary experiment, he took pus from cowpox blisters on the hand of milkmaid Sarah Nelmes and scratched it into the arms of James Phipps, the eight-year-old son of his gardener. James developed a fever and didn't feel well for a short time, but he soon recovered. When Jenner then injected him with pus from actual smallpox lesions, the boy remained disease-free.

Jenner published three papers on the subject and thus became the father of vaccination—the fundamental weapon in the armament of public health. And it began with careful observation.

John Snow, an English physician born in 1813, is considered the patron saint of epidemiology and public health. A member of the Royal College of Surgeons, Snow was a pioneer in the safe administration of anesthesia and administered chloroform to Queen Victoria during the birth of her last two children, in 1853 and 1857.

In those times, London suffered every few years from outbreaks of cholera, which sickened, killed, and spread fear throughout the metropolitan area. The prevailing belief in the medical community was that the outbreaks were caused by "miasma," or bad air. Snow was skeptical and published his doubts in an 1849 paper titled "On the Mode of Communication of Cholera." At the time, microbiology was in its infancy and the bacterium that caused cholera had not yet been discovered. That discovery would occur over the course of a series of studies and publications by Filippo Pacini, an Italian physician, between 1854 and 1865.

The outbreak of August 1854 was the worst in memory, and in some parts of London the mortality rate exceeded 10 percent. One of the most severely hit precincts was Soho, an area of the West End bordered by Oxford and Regent Streets that had seen a large influx of immigrants and the poor and had inadequate sanitation and virtually no sewer facilities.

Snow realized that the largest cluster of cases appeared to be concentrated in a two-block-long thoroughfare in the middle of Soho, near Regent (now Oxford) Circus and along Broad (now Broadwick) Street. He began recording these clusters by blackening out the buildings in which the residents lived on a London map. With the help of Reverend Henry Whitehead, assistant curate of St. Luke's Church and, at the time, a believer in the miasma theory, Snow then went to the homes of the afflicted and asked them about their personal habits and whereabouts in the days before they became ill.

Through this method of shoe-leather epidemiology, Snow came up with an astonishing observation. Nearly all of the victims had taken water from the Broad Street pump. What's more, of the ten deaths mapped closer to another pump, five of the victims had still used Broad Street because they preferred the water. In three other cases, the dead were children who attended school near Broad Street.

Snow viewed samples of the pump water under his microscope and subjected them to chemical analysis. The results were inconclusive. But he was by then so convinced of the association that on the evening of September 7, he went before the Board of Guardians of St. James's Parish, detailed his statistics, and requested that they remove the pump handle, rendering the pump inoperative.

The next day that is just what they did. Though cholera was already waning as many fearful Londoners fled the city, the shutdown of the Broad Street pump effectively ended the outbreak.

Unfortunately, after the cholera crisis was over, government officials gave in to local residents who wanted their well back and replaced the pump handle. It was only in 1866, when a similar cholera outbreak associated with drinking water from another contaminated well occurred, that the Broad Street pump was permanently closed.

Today, the John Snow pub, on the corner of Broadwick and Lexington Streets, is a place of pilgrimage for any epidemiologist or public health officer visiting London. I have been there many times and shared a pint or two. Each time I visit this landmark, I'm reminded that even though scientific research had not yet established the cause of cholera, the basic methods employed by Dr. Snow remain the foundation of epidemiological investigation to this day.

Snow's work was clearly an important milestone in epidemiology and public health practice. But I believe the honor of being considered the father of modern public health could go to Nikola Tesla.

Tesla was the Serbian engineer credited with inventing the alternating-current induction motor and widely applying the use of electricity. The advent of electricity brought about quantum leaps in public health and infectious disease control. With electricity and water pumps, safe water supplies could be realized throughout the world. And with running water, effective sewer systems could be put into place. Electricity also brought us refrigeration, the ability to pasteurize milk, vaccine manufacturing, and air conditioning to keep mosquitoes out of our homes and places of work. It revolutionized medical practice through the invention of X-ray and other imaging technology, diagnostic equipment, mechanical ventilators, and more.

In 1900, the average life expectancy in the United States was forty-eight years. By 2000, just one hundred years later, it was seventy-seven. For every three days we lived in the twentieth century we gained a day of life expectancy. Consider that in light of the fact that early humans in the form of *Homo erectus* emerged 2.4 million years ago, and it took us until 1900 to achieve our forty-eight-year life expectancy. That means it took 80,000 generations to reach the 1900-era life expectancy, and only about 4 to reach our current level. With clean water, sewer systems, safer food, pasteurized milk, and vaccines, we made historic advances in eliminating the diseases that killed children, who are particularly vulnerable to the illnesses related to these environmental conditions.

But lest we start congratulating ourselves too enthusiastically for our progress, as we shall see, the challenges we face going forward are, if anything, even greater than those we faced in the past.

CHAPTER 3

White Coats and Worn Shoes

> *A physician is obligated to consider more than a diseased organ, more than even the whole man — he must view the man in his world.*
>
> — Harvey Cushing, MD

If the white coat is the symbol of hospital- and lab-based medical science, the bottom of a shoe with a hole in the sole is the symbol of the field epidemiologist. In fact, it is the emblem of the Epidemic Intelligence Service, whose motto is "Shoe Leather Epidemiology." Like crime investigation, effective public health requires both the lab personnel and the detectives out at the scene.

My work with toxic shock syndrome (TSS) — which had brought me to the CDC that day in 1981 — turned out to be a classic medical detective story, and one with a surprise ending. It also provided me with a number of career-defining object lessons I have never forgotten.

The term "toxic shock syndrome" was coined in 1978 by Dr. Jim Todd, chief of Pediatric Infectious Disease at Children's Hospital in Denver. For the previous three years he had seen sporadic cases of boys and girls, ages eight to seventeen, presenting with high fever, low blood pressure, rash, fatigue, and sometimes confusion. The first case he saw, a fifteen-year-old boy, was initially diagnosed with scarlet fever, but Jim thought the symptoms seemed much more severe than he would have expected from that condition. Several more cases in the next couple of years were worked up, and although *Staphylococcus aureus* bacteria was detected in the patients' mucosal linings, such as the throat and mouth, none could be isolated in their blood, cerebrospinal fluid, or urine. Based on the severe effects throughout the body, though, Jim and his team suspected a toxin, or bacterial poison, must be involved. One of his young patients had not survived. Lab analysis confirmed enterotoxin type B in blood samples. This toxin is produced by *S. aureus* bacteria.

They published their initial paper in the British medical journal *Lancet*—to more than the usual amount of skepticism from the health community. But Jim's prescient work would serve as a critical first clue and early road map in understanding this apparent new collision between disease-causing microbes and humans.

Without warning, in the spring of 1980, cases of a TSS-like illness began showing up, primarily in Minnesota, Wisconsin, and Utah. Later we would learn that the state-by-state number of cases was largely a result of which states had health departments that were actively looking for TSS cases once the initial alarm had been set off. However, in all three states, nearly all of those afflicted were teenage girls and women in their early twenties. I had been in regular contact during this time with my close colleague and friend Dr. Jeffrey Davis, the state epidemiologist at the Wisconsin Division of Health, about the cases in our two states. Of the twelve cases in the two states, all were young women, eleven of whom had their menstrual periods at the time of their illness onset. Many of the cases were critically ill for up to several weeks; fortunately none had died at this point. Our initial findings did support that TSS was primarily occurring in young menstruating women, but we could not explain the magnitude of the risk, why it was happening, and what to do to stop new cases. We contacted the CDC and they asked other states to start looking for cases.

On May 23, the CDC published an article in *MMWR* describing fifty-five TSS cases in Wisconsin and Utah, forty in which a menstrual history had been obtained; thirty-eight of those—or 95 percent—had illness within five days following onset of menses. The media now started to pay attention.

On June 27, a second *MMWR* report summarized the results of a case-control study that included fifty-two cases—many of which were included in the May 23 report—and fifty-two age- and gender-matched controls. This is a type of epidemiologic investigation where we interview the cases—or the case's family members if the case is too sick or has died—using a comprehensive questionnaire to systematically learn about every possible relevant factor in the case's life that could have played a role in her illness. Then we identify "control" participants: people who are closely matched with the case individuals, for example, by age, gender, and residence, but have not been ill. We interview them using the same questionnaire. Our analysis compares the frequency of factors present among the cases and controls to determine if there are differences that can help us explain why the cases became ill.

That analysis found a statistically significant association between tampon use and TSS; in other words, the difference in tampon use between the cases and controls was very unlikely to happen by chance alone, with a much higher number of cases using tampons compared to controls.

Members of the media and some public health officials began to speculate that the recent high-visibility national rollout of Procter & Gamble's Rely brand of tampons coincided with the increase in TSS cases, though the studies to date had not documented this finding. This media coverage would be significant over the next several months in influencing the results of subsequent epidemiologic studies.

Shortly after the June report, Jeff and I decided to collaborate on a case-control study to figure out why there was this sudden increase in TSS cases associated with menses, and the exact role that tampons and any infectious agent might be playing in this emerging public health concern. We invited the Iowa State Department of Health to participate in the study to help us more quickly identify cases. In our business, outbreaks are defined as a marked increase in cases of a disease, usually in a specific geographic area and over a limited time period.

For whatever reason, we were in the midst of an outbreak of TSS.

Our effort would become known as the Tri-State Toxic Shock Syndrome Study (TTSSS). For our study we had highly trained female investigators do the interviews in private, because they had to ask these young girls highly personal and potentially embarrassing questions. For example, we asked for detailed information about their sexual histories and the use of tampons and pads during their periods. Despite these sensitive questions, every control candidate we contacted agreed to participate. They were the real heroes in our study and helped us save many lives.

Most of the cases we studied had occurred in the previous six months, but we did find some that had occurred even several years before but that hadn't been recognized as TSS. In all three states, we systematically searched in all our hospitals to make sure we had every likely TSS case in women included in our study, even if there was no report of menses or tampon use.

In early September, I experienced one of the lowest and most challenging moments of my career as I observed a sixteen-year-old girl lying in her hospital bed, soon to die from TSS. She was surrounded by her family as she received state-of-the-art supportive medical care. But nothing worked. I can't even say what she had looked like before her illness; now she displayed the extensive classic TSS red rash on her face, hands, and feet. By the time I saw her, her face, arms, and legs were tremendously swollen, making her almost unrecognizable even to those who knew her. The swelling, or edema, is caused by what is known as third spacing—a condition where large volumes of fluid normally in the blood vessels and arteries leak into the patient's soft tissue. This degree of shock, which occurs when there is inadequate fluid circulating in the arteries and veins, is very difficult to reverse. As a result, this young girl's body had gone into multiorgan failure as it struggled futilely to maintain blood pressure. To this day, I have trouble expressing the utter helplessness that we all felt at not being able to do more for her.

As I spoke to her grief-stricken parents, all I could offer was my profound sympathy and a promise that we would get to the bottom of this; that their tragedy would help prevent this from happening to other young women. My daughter Erin—now a physician specializing in neonatology—was two at the time, and as I thought about her growing up, all of a father's protective instinct for his children came flooding over me.

On Friday, September 19, the CDC published in the *MMWR* the results of what was known as the CDC-2 study. It included fifty female TSS cases and 150 female controls. The cases all had onset of their illness in July and August and were reported by a number of states to the CDC; no Minnesota or Wisconsin cases were included. The study again found that tampon use was a significant risk for developing TSS and for the first time found that cases had a 7.7 times higher risk of developing TSS using Rely brand tampons versus other brands. Overall, 71 percent of cases used Rely tampons, but only 29 percent of controls used Rely.

Rely had been developed in direct response to consumer demand. For years, women had been asking for a tampon that could absorb much more menstrual flow and prevent accidental leakage. By the early 1970s, the paper industry had created highly absorbent polymers that could retain twenty times their own weight of fluid. An obvious application was disposable diapers. Procter & Gamble borrowed from its disposable diaper technology to design a tampon that increased fluid capacity from five- to tenfold. Though other companies put out their own competing high-capacity tampons, P&G, using its marketing genius, captured more than 70 percent of the high-absorbency market.

The afternoon before the *MMWR* publication I received a call from an associate director of the Food and Drug Administration (FDA) regarding the pending public release of the CDC study the next day. FDA commissioner Dr. Jere Goyan and his staff had just been briefed on the study results and the Rely tampon connection. Jere was aware of our ongoing epidemiologic studies in Minnesota and Wisconsin and the concern we had expressed in conference calls with federal public health officials about the CDC study results. He requested that Jeff Davis and I fly to Washington to brief him on our ongoing case-control study, which was showing that Rely tampon use was reported in only about half of our cases, suggesting it was not the only problem product. This issue was front and center for the FDA since they regulate the safety and effectiveness of medical devices, including tampons. I agreed to fly to Washington early the next morning, in time for our afternoon meeting. This was the first time I'd ever flown anywhere on just hours' notice, but I would do so many times in the years ahead.

The meeting at the FDA brought no consensus on the meaning of the CDC study results. I flew back to Minneapolis that night and was met with an urgent message asking me to call the senior Procter & Gamble executive overseeing the tampon business. P&G officials had been briefed by the CDC on its study findings earlier in the week. They had lots of questions and had gotten few answers. After a highly successful national launch of Rely over the past year, the officials were now pondering the possibility that their product was killing young women.

I was asked if I would attend a Scientific Advisory Group (SAG) meeting hosted by Procter & Gamble at the O'Hare Airport Hilton on Saturday afternoon and Sunday morning. SAG meetings are not uncommon in the business world, but they hardly ever happen on such an urgent basis. The members of the SAG are typically scientists from outside the company who can provide an objective assessment of what the latest science is saying about the topic at hand. This SAG represented the collective scientific think tank on TSS, although no one from the CDC was invited. I knew I had to go to Chicago despite plans for a long-planned family event on Saturday night. None of the SAG members received payment, just travel reimbursement.

The SAG was chaired by Jim Todd, the original TSS investigator, and his skills as the wise and seasoned sage were evident from the first moment. Jim would provide this same leadership in other forums for many months ahead as we worked to untangle the mystery.

We met late into Saturday evening, going over every piece of information, data, or evidence we had from the current TSS epidemiologic and microbiology studies and any other information that might give us some answers. Sunday morning we summarized our six-plus hours of deliberations. Unfortunately, we had many more questions than answers. Late Sunday morning, a P&G corporate jet from Cincinnati arrived at O'Hare carrying a number of the most senior executives, including CEO Ed Harness. They joined us in our conference room, sitting on one side of the large table. After brief introductions, Jim summed up our findings. Was Rely involved in some way with these TSS cases? The answer was a clear and compelling yes, but how and why was unknown. I continued to push our study's conclusion that it wasn't just Rely, so we shouldn't consider the problem over and behind us.

I will never forget Harness looking at the SAG members and asking, "Tomorrow, can I tell the women working at Procter &

Gamble it is safe to use Rely tampons, or tell the men that they are safe for their wives and daughters to use?"

I looked at Mr. Harness and said simply, "No."

That afternoon, I remember sitting on the short flight back to Minneapolis, realizing Rely would almost assuredly be going off the market the next day. I had learned another career-defining lesson: Most companies are good corporate citizens and will do everything they can to resolve problems if they have evidence that their product is the culprit. P&G had delivered a product to market without any reason to believe they were putting anyone in harm's way. I had no doubt that Ed Harness's decision would be based not on some financial calculation, but on whether the women closest to him could safely use the product.

The TSS/Rely story exploded that September 19 weekend and remained in the headlines for months. The national media played on every young woman's fears for her personal safety. By the end of 1980, LexisNexis, one of the primary companies tracking media coverage in the United States, determined that it was the third-biggest news story of the year, trailing only the presidential election and the Iranian hostage crisis. Coverage of the CDC study brought in almost 900 case reports, enough to reach national epidemic proportions. Ninety-one percent of them were associated with menstruation, and the distinct majority of them involved the use of Rely tampons. Procter & Gamble did, indeed, remove their product from the market the day after the SAG meeting, just a year after its heavily advertised national rollout.

The CDC's public message was that Rely brand tampons were responsible for the outbreak, and with their removal from the marketplace, the threat had now been eliminated.

Rely consisted of polyester foam and a chemical called cross-linked carboxymethylcellulose, along with a coating called a surfactant. Surfactants are compounds that lower the surface tension between two liquids or between a liquid and a solid and make it possible for them to more easily blend together.

Our TTSSS investigative team never dismissed a problem with Rely for one minute. But as far as we were concerned in the Midwest, where the initial cases had presented, simply an association with a particular brand of tampons was not enough. There had to be follow-up studies to get closer to a more complete answer. This is where the TTSSS became critical. We included all cases from October 1, 1979, through September 19, 1980, in the three states. There were eighty in all, which we age- and gender-matched with 160 controls. We stopped enrolling new cases on September 19 because the CDC study reports all but guaranteed a bias toward the selective diagnosis and reporting of cases that used Rely tampons going forward.

By the time the study was fully under way, I probably knew more about tampons than 99.999 percent of the male population, more than it had ever occurred to me I'd have to wrap my brain around. I could identify all twenty-one brands and styles sold in America, both right out of the package and after use. You never know what you might face when you venture into the world of investigative epidemiology, and you have to develop a certain degree of scientific detachment. At the same time, I kept thinking about the effect this epidemic was having on millions of women and their families throughout the country. It seemed a cruel irony that this wave of illness and death had involved a product called Rely.

What we found in our study didn't really surprise us. As we wrote in the summary of our paper that would be published in the April 1982 issue of the *Journal of Infectious Diseases,* "By multiple logistic regression analysis, the risk of TSS was more closely associated with tampon fluid capacity (absorbency) than with the use of all tampon brands."

For those who used the lowest-absorbency tampon regardless of brand, there was about a 3.5-fold increase in the chance of developing TSS versus never using a tampon. For those who used the highest-absorbency tampons—of any brand—there was a 10.4-fold increase in developing TSS. However, we did find that Rely users still had a 2.9-fold increased risk compared to users of other brands. While we had evidence there was something special about the risk of using Rely tampons, the real driver in the chance of developing TSS was the fluid capacity of the tampon a woman chose to use. And the TTSSS finding virtually predicted what would happen with cases in our states in the months following the removal of Rely from the market.

The number of cases of young women with TSS did not change much; actually, it rose slightly. What happened instead was that those who came down with toxic shock syndrome were now mainly users of Tampax Super Plus brand high-absorbency tampons and a few other competing products.

Not surprisingly, young women continued to use high-absorbency tampons because no one warned them about the real risk factor. And the prime beneficiary of P&G's decision to withdraw Rely tampons? Tampax. Suddenly it owned more than 70 percent of the high-absorbency market. It then became abundantly clear in states with active efforts to find TSS cases that the problem could not just be Rely; it had to be the use of high-absorbency tampons of any kind.

What that meant was that data from the previous CDC study had been subjected to biased, selected national case reporting due to media coverage of the role of Rely tampons in causing TSS, and was completely misinterpreted. We eventually determined that the key factor in the development of TSS and the relationship between fluid capacity was the increased release of oxygen in the vagina with high-absorbency tampons and the presence of *S. aureus* bacteria. As the menstrual fluid was absorbed into the highly absorbent material, oxygen was displaced into the vagina. The higher the absorbency, the more oxygen was released.

The rise in TSS cases happened to coincide with a new strain of the *S. aureus* bacterium that was a very effective producer of the TSS toxin. But more important, the materials of the highly absorbent tampons released a greater amount of oxygen into the vagina, which should be an anaerobic—oxygen-free—environment. With no oxygen, there is no TSS toxin produced. But this excessive oxygen transformed the bacteria into microscopic toxin-producing factories. Once produced, these toxins were absorbed through the vaginal mucosa—the membrane lining the vaginal walls—and straight into the bloodstream.

Subsequent work over the next several years by Dr. Patrick Schlievert, a microbiologist and internationally recognized expert on staph and strep toxins who had recently left the University of Minnesota for UCLA, and two other research groups demonstrated that the surfactant used for coating Rely tampons—known as pluronic L-92—also increased toxin production, and the surfactants the other companies used did not. Now the TTSSS case-control study results made perfect sense.

Ironically, shortly after the CDC announcement on September 19, the American College of Obstetricians and Gynecologists speculated publicly that it was a personal hygiene issue and recommended that menstruating women change tampons more frequently.

This turned out to be exactly the wrong advice. By telling them to change their high-absorbency tampons more frequently, the college was putting women at higher, rather than lower, risk. The more frequently a woman changed her high-absorbency tampon, the more oxygen she introduced into her vagina. Another lesson I learned from my experience investigating TSS is that if you don't know what you're talking about, then don't talk, or at least say you don't know. Yes, women wanted and needed sound and timely expert advice about the use of tampons, so it's understandable why the American College of Obstetricians and Gynecologists felt the need to make a statement. But the only real information they had at that point supported not using tampons at all.

The prestigious National Academy of Sciences' Institute of Medicine (IOM, now called the National Academy of Medicine) put together a blue-ribbon committee in 1981 to examine in detail the different findings from the various TSS studies and results of ongoing surveillance in states such as Minnesota. The final IOM report confirmed that our study and disease surveillance were, in their words, the "gold standard." What really mattered was that over the ensuing months, all the tampon manufacturers, reacting to the TTSSS findings, greatly reduced the fluid capacity of their highest-absorbency styles, and cases of TSS dropped dramatically.

The TSS investigation was not only my personal launching pad into the big leagues of epidemiologic investigation and analysis; it also made me realize how easily data can be misinterpreted into flawed science and how important it is to get a number of perspectives on board. And it taught me to make sure you ask the right questions, lest you be led to the wrong answers.

In this case, I'm certain that the wrong conclusions from CDC officials regarding TSS and the continued use of high-absorbency tampons resulted in many more women becoming seriously ill and even dying. To this day, I still wonder how many of the TSS-related deaths could have been prevented had the findings of the TTSSS been supported by the CDC and promoted to the public before tampon fluid capacity was reduced by the manufacturers several years later.

Not every outbreak has to have deadly consequences to have a significant effect on a community or to provide important lessons in public health.

It was early in the afternoon of July 10, 1984, that I got a call from Dr. Ron Sorenson, an internal medicine doctor at the Brainerd Medical Center. Ron informed me that at least thirty patients had been seen at his hospital since March with unrelenting chronic diarrhea; none had yet recovered. Despite the fact that eight had been referred for further evaluation at the Mayo Clinic, the University of Minnesota Hospitals, and the Minneapolis Veterans Administration hospital, no cause could be identified.

Located about two hours' drive north of the Twin Cities, the beautiful lake country of Brainerd, Minnesota, has been the go-to place for summer fun at one of the hundreds of crystal clear lakes. But to this day, the mental image of Brainerd takes on a double meaning for me: lakes and diarrhea, and I mean a lot of both.

No doctor or clinical laboratory director had thought to report these cases to the Minnesota Department of Health because, basically, no one knew what illness to report. To make matters more complicated, each of the eight patients seen at our state's leading medical centers was given a different diagnosis with a generalized label like irritable bowel syndrome, nonspecific colitis, or chronic diarrhea of unknown etiology. Two of these patients were seen by the same expert physician team just two months apart, but despite identical illnesses, they came away with different diagnoses. The physicians hadn't made the connection that both patients were from Brainerd and both had become suddenly ill at about the same time.

No one wants to talk about diarrhea; it's almost as embarrassing as having lice. So members of the Brainerd community were unaware of the illnesses that were occurring around them. And because the Brainerd Medical Center had thirty-six physicians for a community of 14,000 people, it wasn't until early July that the doctors made the connection that something unusual was happening.

Because I am an epidemiologist, my interest is always piqued when someone reports a regional cluster of similar illnesses that appears to be out of the ordinary. It was clear to me on that first call with Ron that the chances of seeing thirty-plus patients in the past five months with new onset of severe, chronic diarrhea in a town the size of Brainerd, and all presenting to one medical center, was like winning—or maybe losing—the lottery.

During our call, Ron provided the details of one of the patients, whom I'll call John. John was a healthy seventy-seven-year-old man who had developed sudden onset of watery diarrhea. He had few other symptoms; no nausea, vomiting, cramps, or fever. For the next month he had ten to twenty bowel movements a day and lost more than twenty pounds. After numerous stool samples came back negative for the typical causes of infectious diarrhea, he was hospitalized as one of the eight patients noted above. The only remarkable finding was that a colonoscopy showed inflammation of his colon. He was diagnosed as having nonspecific colitis of unknown etiology. He was treated with several antibiotics but there was no change in his symptoms.

John's social life and everyday activities suffered, as he could not be far from his bathroom. Over the course of the next year, his diarrhea continued with only a slight reduction in the number of episodes, though he realized he could eat more food without substantially changing his bathroom routine. As a result, he gained back some of the weight he'd lost. During the second year he noticed that he was having fewer and fewer diarrheal episodes. And by 550 days following his first symptoms, the frequency and volume of his stools returned to normal.

Within minutes of Ron's call, I gathered our senior infectious disease epidemiology and laboratory team at the Minnesota Department of Health. A group of us would head to Brainerd that night to begin our investigation.

I strongly suspected an infectious disease microbe was to blame for this outbreak because of the sudden onset of illness in so many individuals. So we called our colleagues at the CDC foodborne division and shared what we'd learned so far and requested their lab support. They had two of their staff join the investigation.

The new CDC EIS officer, who was just learning the ropes of outbreak investigations and who would arrive on a plane from Atlanta the next day, would become my professional soul mate. Dr. Kristine MacDonald—now Kristine Moore—provided invaluable leadership during this investigation. When she completed her EIS stint, she took a position at the Minnesota Department of Health as the assistant state epidemiologist. We have been a synergistic team ever since, and as I tell my students frequently, epidemiology is a team sport. I couldn't have accomplished half of what I did without Kris as my professional partner.

Kris recalls, "The biggest issue was trying to determine the etiologic agent and how people were being exposed to that agent. Then: How large was the affected cohort? How much of the community was impacted?"

The first thing we had to do when we arrived in Brainerd that night was to pore over the records of patients the clinic had seen for diarrhea in the past six months. If this was a real outbreak, we should be able to pinpoint when these cases started appearing. We also used the clinical information on the patients who'd had extensive medical workups to start to develop a case definition.

We defined a case as someone with diarrhea of unknown etiology lasting four or more weeks. As we learned more about these cases and the outbreak in general over the weeks that followed, this definition held up as being both sensitive, picking up all cases, and specific, not including any cases of diarrheal illness that were due to some other cause. Since we had not identified an infectious or chemical reason for the illness, we had to use a combination of clinical findings to define the outbreak-associated cases and distinguish them from cases with known causes like Crohn's disease or colon cancer.

We quickly reviewed the thirty-plus cases Ron had described to us over the phone. We included the first twenty-three that met our case definition and had onset of illness between April and June 1984. We also identified forty-six gender- and age-matched controls who had not had diarrhea during this same time—sixty-nine in total. We asked about everything imaginable that might happen in a person's life in a given month. In particular, we asked about everything they had consumed in the month before, including medications.

Kris took over as lead of the clinical and microbiologic aspects of the investigation while I concentrated on the epidemiology.

We hit pay dirt almost immediately. The first three cases, none of whom knew one another, reported that they routinely consumed raw milk sold by a local dairy located just outside the Brainerd city limits. We knew we had to be very careful while conducting our subsequent interviews so we didn't lead interviewees into recalling a history of raw-milk consumption and biasing the results, but this lead was nonetheless a gold mine.

The critical relationship between illness and raw-milk consumption quickly became clear and compelling. The case-control study found raw-milk consumption the only one that stood out among the hundreds of factors we considered. In fact, cases were more than twenty-eight times more likely to have consumed raw milk from the local dairy than the controls.

In 1864, Louis Pasteur discovered that heating beer and wine to temperatures less than boiling for variable time periods was enough to kill most bacteria. This process prevented these beverages from spoiling while not altering their quality or taste. Today, the process of pasteurization is used widely in the dairy and food industries for microbial control, thus ensuring the safety and preservation of milk.

Raw milk, which some people still consider to be healthier and more nutritious, is not pasteurized. Before the age of routine pasteurization, many people, especially children, fell prey to a number of dangerous diseases as a result.

So we had an answer to the *why* in Brainerd. But there was still a lot we didn't know. *What* was causing this illness? Was it an infectious disease, and if so, were the cows infected? Could others who hadn't drunk the raw milk become infected from the ill cases? Were there any treatments that could reduce the symptoms or even cure the illness? Was this just the tip of the iceberg?

Investigative priority number one: Stop the outbreak. After confirming the local dairy's milk as the source of the microbe or chemical causing the outbreak, our first act of business was to make sure no additional milk was sold from that farm. The farmer quickly understood the extensive body of evidence we had implicating his milk with the diarrheal illness. He agreed not to sell his raw milk to any source unless it was going straight into a plant for pasteurization. Observation and the use of epidemiologic studies allowed us to "pull the pump handle," even though we hadn't yet discovered the specific cause of the outbreak. After the sale of the raw milk stopped, so did new cases.

Eventually, we confirmed 122 cases of chronic diarrhea among the drinkers of raw milk from that specific dairy. The first had onset in December 1983 and the last in July 1984. Together, the Minnesota Department of Health and the CDC threw every lab resource we could into analyzing this outbreak, yet we could not identify even a suspect infectious virus, bacteria, parasite, or chemical cause in either the human cases or the cattle herd at the dairy. And it wasn't because we didn't have lots of fresh specimens.

After much discussion among my colleagues at the Minnesota Department of Health, the CDC, and staff at the Brainerd Medical Center, we decided this disease needed a name. We designated it "Brainerd diarrhea," following the then current practice of using geographical names, such as Lyme (Connecticut) disease and Norwalk (Ohio) virus. Brainerd diarrhea is the officially recognized name for this condition in the medical literature.

"Despite a really extensive, elegant investigation with the most current testing methods, we never did find the etiologic agent," says Kris. "But we really put the condition on the map."

Through extensive efforts to uncover previously unreported outbreaks or single cases, we found a similar clinical illness among raw-milk drinkers in Minnesota (1978–79 and 1984), Oregon (1980), Wisconsin (1981–83), Idaho (1982), Massachusetts (1984), and South Carolina (1984). In addition, at least ten outbreaks have occurred since Brainerd, including large ones in Illinois and Texas. In each one, either raw milk or contaminated water was responsible.

I'm convinced that Brainerd diarrhea is caused by an infectious agent and that we will eventually find it.

As we have seen with HIV/AIDS, toxic shock syndrome, and Brainerd diarrhea, virtually nothing that happens in life is off-limits or irrelevant to the epidemiologist's purview. It ranges from the most intimate and personal aspects of individual biology all the way up to the most public and far-reaching geopolitical clashes.

The lesson the Brainerd experience taught me was: You don't have to have all the answers to have the critical answer. Like John Snow, we can stop or limit the occurrence and impact of infectious diseases without knowing everything about them. I often hear that we can't act on this or that because we don't have all the answers. That's nonsense. We have to be prepared to go into battle with the knowledge and resources we have, beginning with basic observation.

And we can!

In the early days of the 2015–16 Zika outbreak in the Americas, I found myself repeatedly frustrated by scientists and journalists who had never been involved with an actual outbreak investigation declaring that we didn't have proof that the Zika virus caused microcephaly and Guillain-Barré syndrome and thus that all public health recommendations were not based on conclusive evidence. From my experience, I considered the evidence abundant and conclusive, and any delay in responding was an irresponsible and indefensible position.

My colleagues and I have often been criticized by politicians and the media for "making it up as we go along," to which I plead 100 percent guilty. When we are in the midst of pursuing a serious outbreak of unknown origin or scope, we *are* making it up as we go along. Being a public health official leading a serious infectious disease outbreak investigation means you often have to make quick decisions about taking action in order to prevent additional cases and even deaths. The challenge is not to be wrong, because your credibility will be forever challenged if you are.

As Bill Foege puts it, "You have to make adequate decisions based on inadequate information." That is the very nature of epidemiological investigation. The important thing is for the public to understand this and have confidence that competent and dedicated men and women are on it: that they are telling you exactly what they know and don't know, and what they are doing to "remove the pump handle."

CHAPTER 4

The Threat Matrix

> *Like Abraham Lincoln, I am a firm believer in the people. If given the truth, they can be depended upon to meet any national crisis. The great point is to bring them the real facts.*
>
> — General Douglas MacArthur, 1944

A threat matrix is a graph that shows us what we should be worrying about. In epidemiology, we have several ways of constructing a threat matrix.

One has a vertical axis that measures impact risk and a horizontal axis that tracks emergence risk. So a potential pathogen that would have a great impact but is not likely to emerge would occupy a lower-risk quadrant than a high-impact, high-risk-of-emergence pathogen.

A matrix that I consider equally important has a horizontal axis tracking the potential severity of the pathogenic event and a vertical axis measuring the degree of preparedness. Using this threat matrix, we can determine our likelihood of meeting the threat, whatever it is. But as simple as that sounds, it involves a lot of variables.

Public health science is based on statistics and probabilities. But we as a population don't think in those terms. If we did, no one would ever buy a lottery ticket. Rather, we think emotionally, especially about things like disease and death. Therefore, our personal threat matrix is not likely to conform to the qualitatively and quantitatively based ones we just mentioned.

For example, we all know intellectually that mile for traveled mile, airplanes are far safer than automobiles. Yet those of us who are afraid of flying will get in our cars every day without giving a second thought to the risks of the road. In the same way, we will tolerate 40,000 or so highway deaths a year in the United States, but when the I-35W bridge over the Mississippi River collapsed not far from my Minneapolis office in 2007, killing thirteen people, we were all shocked and outraged. We hadn't assimilated bridge and tunnel failure into our personal threat matrices.

As a result of the nearly 3,000 civilian lives lost on 9/11, the United States embarked on a multitrillion-dollar challenge to the threat of terrorism that reorganized much of government and resulted in profound changes in the way we live, travel, defend ourselves, engage in foreign conflict, and go about our daily lives. Certainly, this effort may have prevented terrorist incidents or discouraged would-be terrorists. And certainly, I understand that the terror factor far outstrips the simple number of deaths. But it is hard to argue that the response has been a proportionate one compared to our responses to other threats we face.

We need to have a real-world assessment of infectious disease risk.

In a 2015 TED Talk, Bill Gates asserted, "If anything kills over 10 million people in the next few decades, it's most likely to be a highly infectious virus rather than a war. Not missiles, but microbes. Now, part of the reason for this is that we've invested a huge amount in nuclear deterrents. But we've actually invested very little in a system to stop an epidemic. We're not ready for the next epidemic."

In public health, as in other areas of life, you can't plan for everything. We can look at examples in disaster management and business continuity planning. After the terrorist attacks of 9/11, a number of large corporations in New York City decided they'd better have power in case this kind of horror happened again. So they placed emergency generators in the basements of their buildings, well protected from potential air attack. But they hadn't planned for an event like Hurricane Sandy in October 2012, which flooded lower Manhattan and even parts of the New York City Subway.

What we can do as a society, though, is plan generally for disaster: for suspension of power, discontinuity of service, medical emergencies when resources might not be available, and self-sustaining survival before help arrives. As President Dwight D. Eisenhower said, "In preparing for battle, I have always found that plans are useless, but planning is indispensable."

In the 1990s, coauthor Mark Olshaker was researching and writing an IMAX film on "big weather"—hurricanes, tornadoes, and monsoons. While visiting the National Hurricane Center in Miami, Florida, with his producer-director Greg MacGillivray, Mark asked the center's distinguished director, Bob Sheets, what the worst nightmare was for a meteorologist in his position.

"That's easy," Sheets replied. "Category five hurricane—direct hit on New Orleans."

On August 29, 2005, Hurricane Katrina hit New Orleans. By the time it reached land, it had been downgraded to a Category 3 storm. It still managed to kill 1,577 people in Louisiana alone, displace many thousands more, and completely disrupt the life of that great American city, in the process becoming the costliest natural disaster in American history.

Despite the fact that Sheets's warning was common knowledge in scientific and emergency management communities, no one had sufficiently prepared for such a disaster. A missed opportunity to take proactive measures? This is what we have been facing in the public health field as far as preparing for infectious diseases in the twenty-first century: one missed opportunity after another.

* * *

There are only four events that truly have the power to negatively affect the entire planet. One is all-out thermonuclear war. Another is an asteroid striking earth. The third is global climate change. And the fourth is infectious disease.

Thermonuclear war speaks for itself, and we can only hope world leaders are sufficiently rational and enlightened to avoid that catastrophe. Fortunately, terrorists do not yet have the ability to inflict such horror, even if they happen to possess or seize a single nuclear device.

An asteroid strike is highly unlikely, and anyway, there's not much we can do about it.

We have already emitted so much greenhouse gas that climate change is an established fact. Even at current levels, it will result in a worldwide crisis that will unfold over several decades or more. But during that time, we can develop plans to deal with coastal flooding, with the impact of too much or too little rainfall, and with the effects that changing temperatures will have on animal, plant, and insect populations.

Infectious diseases in the twenty-first century have the greatest potential of these four events, I believe, to give rise to a sudden crisis that would involve the whole world at the same time—a pandemic, or worldwide epidemic.

At this point, our collective major concern should be pandemic influenza, though as we have seen from HIV/AIDS, other microbial agents can emerge unexpectedly.

Unlike Superstorm Sandy, Hurricane Katrina, the 1989 Loma Prieta earthquake, a tornado, or any other natural disaster that wreaks massive destruction and then ends quickly so that recovery can begin, a pandemic spreads around the world and lasts for an extended period of time. It does not hit just one locale, leaving all others with the ability to come to its aid. A pandemic hits many locales simultaneously, all of them needing emergency assistance. It has a rolling effect as it hits first individuals, then civil authority, then business, then interstate or international commerce or both. The effects are immediate and devastating, the consequences long-term.

When everyone is involved in a pandemic, no one has extra help or supplies or food or medicine to send around, unless there was sufficient planning. There is a naïve belief that the kinds of supplies we need to respond to a pandemic, such as medical products, drugs, vaccines, and N95 respirators—commonly known as face masks—will be a click away on the Internet. Not so.

Today, we live in a just-in-time-delivery economy where virtually nothing is warehoused for future sales, let alone stockpiled for a crisis situation. Not even the parts and components necessary to manufacture these critical supplies are warehoused or stockpiled. When a rolling global pandemic takes its toll on the working population of a city in Asia, for example, the products and supplies that come from that city—and perhaps nowhere else—that we need to respond to a rapidly growing pandemic will not be available. No amount of money can buy something that doesn't exist. This is why the recently created World Bank Pandemic Emergency Financing Facility fund, which is intended to provide global financing for responding to a pandemic, will not work for a global emergency.

If a major pandemic hits, no matter where we live, we will be largely on our own. One case of Ebola sent shock waves throughout Dallas, Texas, in 2015. What if Dallas and cities all around the world were experiencing thousands of cases at the same time?

Even though it is an "act of nature," a pandemic is much closer to war than any other natural disaster. As in war, in a pandemic, there is greater destruction day by day, with no opportunity for recovery.

Even if an outbreak does not spread beyond a region, it can still be devastating. I call these "outbreaks of critical regional importance." The 2003 SARS (severe acute respiratory syndrome) outbreak was exactly that. It was limited to a few cities around the world such as Hong Kong and—through airplane travel—Toronto. Nonetheless, it caused death and tremendous human suffering in those areas and had a severe economic impact.

Early in 2015, I addressed a conference at the Institute of Medicine in Washington in which I predicted that the coronavirus MERS—Middle East respiratory syndrome, a kissing cousin of SARS—was bound to cause serious outbreaks outside of the Arabian Peninsula in the very near future. I couldn't predict where, of course, but I knew it would happen.

Sure enough, weeks later it turned up in Seoul, South Korea, one of the most technologically sophisticated cities in the Pacific Rim. One "superspreading" individual shut down Samsung Medical Center, one of the most advanced hospitals in the world, and created a governmental crisis. Can you imagine the impact of a single contagious person shutting down Bellevue Hospital, Massachusetts General, Cedars-Sinai, or the Mayo Clinic?

Every time there is a major disease outbreak—Ebola in 2014, MERS in 2015, Zika and yellow fever in 2016—I get calls from media throughout the United States and the world, looking for explanations, guidance, and predictions. I'm generally happy to comply, but I also must admit to a frequent sense of déjà vu in such cases when I think of all the opportunities we've had to take proactive measures that might have prevented, and certainly would have mitigated, whatever situation or crisis is currently before us.

All of the battles against our deadliest enemy are worth fighting, but some have to be fought more quickly and vigorously than others. This is not a question of infectious disease versus chronic, or epidemic versus endemic. It's not even a question of how much of our resources go to medicine and public health versus how much go to antiterrorism. Every death or serious illness from infectious disease is a crisis for the individual patient, his or her family and close friends, and the physician and medical team. But some infectious diseases become crises for regions, nations, or the world, threatening social, political, and economic stability.

Since we can't actively deal with everything, we propose four orders of priority that, we will argue, should lead to nine distinct but interrelated endeavors we collectively call our *Crisis Agenda.*

The first priority is to confront head-on those microbes that cause deadly pandemics, or as we refer to them in our business, pathogens of pandemic potential. They are the deadliest of our deadly enemies. There are only two microbial threats that I believe fit this description. The first is influenza: the one respiratory-transmitted infection that can be spread around the world in short order and strike with lethal force.

The other pathogen of pandemic potential is actually a growing number of virulent microbes that are more insidious in their transmission but will still greatly impact the health of humans and animals around the world. This is the threat of antimicrobial resistance and the very real possibility of moving ever closer to a "postantibiotic era." Imagine a world more like that of our great-grandparents, where deaths due to infectious diseases we now consider treatable are once again commonplace.

The second priority is to prevent high-impact regional outbreaks, such as Ebola and coronavirus infections including MERS, and the possible return of SARS and Zika as well as the other mosquito-borne diseases that continue to have such a devastating impact on the world's poor and that disrupt national economies and governance.

The third priority is to prevent the use of microbes for intentional harm, to prevent the accidental release of a microbe that has been enhanced by scientists to be more easily transmitted, to be more likely to cause death or serious disease, and to be unpreventable by vaccination or treatment with antimicrobial drugs. This priority includes the issues of bioterror and dual-use research of concern (DURC), and research-based gain-of-function research of concern (GOFRC) studies.

DURC essentially refers to scientific research that, based on current understanding, could reasonably be anticipated to be used not only for beneficial purposes, but also to cause harm, either by intentional application or by accident. According to the National Institutes of Health (NIH), "The United States Government's oversight of DURC is aimed at preserving the benefits of life sciences research while minimizing the risk of misuse of the knowledge, information, products, or technologies provided by such research."

GOFRC describes scientific studies or experimentation that increases a pathogen's ability to cause disease, become more transmissible, or make that disease more severe, more difficult to treat, or both.

The fourth priority is to prevent endemic diseases that continue to have a major impact on the world's health, particularly among emerging nations. These include malaria, tuberculosis, diarrheal diseases, and AIDS, which, despite the advances we have made, may be thought of as slow-moving pandemics.

We're going to address these priorities head-on throughout the book, and we will also zero in on the things that truly are worth worrying about. But one thing I want to stress here and now is that it isn't merely a question of science.

Beginning with chapter 9, we have organized the book in ascending order of the Crisis Agenda priorities, concluding with the two that have the ability to substantially alter our everyday lives: antimicrobial resistance and pandemic influenza.

CIDRAP, the organization I founded and head at the University of Minnesota, is an acronym for Center for Infectious Disease Research and Policy. Research *and* policy: Like chocolate and peanut butter, these two have a natural connection. If we approach science without policy, we will accomplish nothing. And if we try to institute policy without good science behind it, we will squander precious time, money, and lives.

CHAPTER 5

The Natural History of Germs

When things get bad enough, then something happens to correct the course. And it's for that reason that I speak about evolution as an error-making and an error-correcting process. And if we can be ever so much better—ever so much slightly better—at error correcting than at error making, then we'll make it.

—Jonas Salk, MD

The comparison between crime detectives and disease detectives holds true on many levels. In this context, we can think about microbes the way we think about people.

We're pretty much constantly surrounded by other people. Most of the time we encounter the same people every day, but we also see some different people every day. Most don't affect our lives one way or another; we simply occupy similar or contiguous space. But there are friends, family, loved ones, and coworkers who do make a positive difference in our lives.

Others who we never meet are still critically important to our everyday lives; we just don't think about it. For example, when was the last time you thought to thank the person who runs the electricity-generating plant a hundred miles from where you live or work and is responsible for keeping your lights on and your grocery store freezers and refrigerators running?

How about the person who drives the delivery truck and makes certain the lifesaving drug your family member desperately needs today is in the hospital pharmacy when it needs to be? These are faces we never see of the people whom we really count on.

Then there are a few mean, dishonest ones or criminals who can impact us in decidedly negative ways. In the most extreme instances, they can actually end our lives.

So it is with microbes. Most don't impact us positively or negatively. Some are essential to the maintenance and quality of our lives, and some are predatory and harmful. What we call criminals in the human sphere we call pathogens in the microbial realm.

Only recently have we begun to realize how we as humans coexist with the global array of microbes—what we call the microbiome. Unfortunately, we still have a largely naïve view of that relationship, often shaped by popular media figures expressing their disgust when someone reports that samples taken from phones or door handles in our offices or homes are flush with germs. This simplistic view is like concluding that the only good plant is a dead one because you don't want weeds in your yard. To understand the potential for pathogens, we need to start at the very beginning of time.

Earth began as molten rock about 4.5 billion years ago. Sometime in the next billion years, one-cell life appeared in the developing oceans of the planet in what is referred to as the primordial soup. There are several theories of how and why these cells appeared; we may never know for sure what really happened. In the 1920s, Soviet biologist Alexander Oparin and British geneticist J. B. S. Haldane proposed theories that ultraviolet radiation provided energy to convert methane, ammonia, and water into organic compounds. As certain molecules combined, they achieved survival advantages.

A more recent theory suggests that simple organic life was

achieved from chemical energy arising from earth's thermal vents. More theories are likely to arise.

What is relevant to our perspective is that for more than 3 billion years, microbes were the only life-form on earth. Microbial evolution allowed them to literally become the reason humans, animals, and plants could exist. They created the oxygen atmosphere we need to breathe and the ability of plants to get the carbon dioxide and the nutrients in the soil they need to grow. These are the foundations of life as we know it today.

Evolution is the force that drives diversity, and evolution is based on stress. The better an organism of any size—bacterium, woolly mammoth, human being, or blue whale—can cope with or adapt to stress, the better its survival chances. There can be major and immediate stressors, such as a large meteor hitting the earth. Most stress, however, occurs over millennia.

For about 3 billion years, all of evolution involved bacteria, single-cell organisms without a nucleus. Inexorably, over a period of eons nearly incomprehensible to the human temporal context, these microbes combined and evolved into all of the plant and animal life-forms that have ever existed on earth.

Without going into all the complex biochemistry of diversity, the important thing for us to remember is that microbes were here before us, have coevolved with us while we humans occupy the earth, and will be here after we are gone. In our superior human mind-set, we think of our species as being largely in control. But to understand the true biologic sense of the power of microbes, we must never forget that we are the ones trying to anticipate and respond to their evolution, not the other way around.

We need many of the current microbes to survive. But some can kill us.

As my friend and colleague Dr. Martin Blaser, professor and director of the Human Microbiome Program at the NYU School of Medicine and one of the most respected infectious disease minds in our business, notes in his enlightening book, *Missing Microbes,* "Bacterial cells are complete, self-contained beings; they can breathe, move, eat, eliminate wastes, defend against enemies, and, most important, reproduce." In sum, Blaser writes, "Without microbes, we could not eat or breathe." So when we lose essential bacteria, we don't do very well.

At the very end of the story until now—the human chapter—we experienced hyperevolutionary bursts. But in spite of our human standing in the modern world, microbes—the microbiome—still outweigh every other component of earth's biomass combined.

There are more microbes in the human gut than there are cells in the entire body, and there are microbes virtually everywhere within us. Yet our personal microbiome accounts for just about three pounds of our total body weight. So for the total of microbes on earth to outweigh all other life-forms, their predominance in our existence is staggering to contemplate.

It's critical that we don't throw the baby out with the bathwater. We need to hold in great scientific reverence those microbes that sustain us as healthy humans, animals, plants, and environment. In fact, we need to further our research and policy agendas that support their survival. It's no different from ensuring the healthy existence of our rain forests to combat climate change.

Now, having laid this all out, we have to understand that we humans and animals start out at a disadvantage. As a species, we reproduce on average about every twenty-five years, the rough definition of a human generation. Microbes, on the other hand, can reproduce about every twenty minutes. By our standards, they are hyperevolutionary. So you can see that in this war, ours is not the dominant or strategic form of renewal.

To further complicate the matter, we are altering the dynamic with pathogens simply through our encounters with them. By venturing into the microbes' homes deep in rain forests, for logging, planting, and hunting for bushmeat; by concentrating large numbers of people together; by breeding millions and millions of pigs and poultry and keeping them in close confines; by overusing and misusing antimicrobial drugs, we humans are forcing microbes to adapt to continual stresses and giving them opportunities nature never did.

Don't we adapt too? Of course we do, but you can see how many microbial generations equal one human generation: about 40 million to 1. It would be as if the Grand Canyon were created by high-pressure water cannons in a day rather than by drip-by-drip erosion over many millennia. Europe suffered a huge depletion of labor, productivity, and social advance in the third decade of the twentieth century, results of the twin devastations of World War I and the 1918 flu pandemic. If we wipe out a correspondingly large chunk of microbes, the strain can recover in about a day.

There are many families and orders of magnitude within earth's microbiome. In order of size and complexity, they include prions, viruses, rickettsia, bacteria, fungi, and parasites. We're going to focus on those microbes that have the potential to kill or seriously hurt us as well as to disrupt the social, economic, and political fabric of the world, or at least major parts of it. As you will note, the viruses dominate in this category. They inflict an enormous hit on humans, animals, plants, and even other microbes such as bacteria.

Viruses are not, strictly speaking, alive, but neither are they inorganic. They exist in a sort of intermediate netherworld, lying in wait until they can hijack the reproductive mechanism of a living cell and get it to churn out copies of the virion by the millions. There is often a host target: a situation where a specific virus may infect only humans or certain animal species. A good example of this is smallpox virus, known as variola. It will infect humans but not animals. On the other hand, there are those viruses that will readily infect both humans and animals, like rabies. Often there also is a high level of organ tropism, meaning the virus tends to infect only certain organs or body parts in a host. For example, the human hepatitis viruses largely infect the liver.

Viruses, like most other microbes and higher orders of life, reproduce according to the dictates of DNA or RNA: the long molecules that make up our chromosomes. Once a virus enters a victim's cell it must reproduce, and this where the importance of viral genetics comes into play. It is far beyond our scope here to get into the complex world of virus replication. Understanding in detail whether an RNA virus is single-stranded or double-stranded, is an RNA (+) sense or (–) sense, or uses a DNA intermediate isn't something you have to understand to determine which viruses should rise to the head of the list of agents of greatest concern for pandemic or critical regional potential.

What is important is that we, as public health scientists, determine which infectious disease microbes can rapidly mutate or change their genetic codes effectively to avoid the host's immune system, vaccines, or drugs, and can even lead to enhanced means of transmission, particularly through the respiratory route. This is why influenza viruses remain the leading candidates for causing a global pandemic.

Sometimes antigenic changes render the microbe less harmful, sometimes more. As we've said, each generational transmission is a roll of the genetic dice.

Individual components of our blood, including B cells and T cells, seek out foreign invaders and use their various mechanisms to envelop or destroy them, or both. They stick around for some period of time, some of them for a lifetime, with the "memory" of the invader, so that if it strikes again, the immune system is ready for it without having to ramp up as much as it did the first time it encountered this invasive agent. This is the concept behind vaccines: introduce an attenuated or dead version of the virus so that the body can build up these defenses before the "real" one hits.

In some instances, the offending microbial agent is merely the trigger; the "bullet" comes from our own bodies. This occurs when the microbe induces an overvigorous response from the immune system, triggering what is known as a cytokine storm. Cytokines are small proteins that alert the appropriate white blood cells to rush to the site of infection and fight the invaders. In a cytokine storm, the continual feedback loop between cytokines and the defensive cells can clog airways and cause organ shutdown. This is what we believe happened with the 1918 flu strain, why it killed so many young and previously healthy people with robust immune systems.

We have factored the means by which microbes replicate into our categories of which microbes to be most concerned about. Microbes that change their antigens or component parts quickly through mutations in their genetic footprint will rate a high-concern score if they also spread via the respiratory route and can more effectively kill those they infect. Developing an effective vaccine for this category of microbes is more challenging but also more critical than developing a vaccine for less deadly microbial forms.

The battle lines are well drawn: the microbes' genetic simplicity and evolutionary swiftness against our intellect, creativity, and collective social and political will. We cannot overwhelm the pathogens, because they so vastly outnumber and outmaneuver us. Our survival depends on outsmarting them.

CHAPTER 6
The New World Order

People are beginning to understand there is nothing in the world so remote that it can't impact you as a person. It's not just diseases. Economists are now beginning to say if we are going to have good markets in Africa, we're going to have to have healthy people in Africa.

— William Foege, MD

For most of human history, infectious disease outbreaks weren't much of a concern compared with the other challenges of staying alive and finding enough to eat. When our ancestors lived in small groups of hunters and gatherers, there wasn't enough of a population concentration to create much of an epidemic. But about 10,000 years ago, with the beginnings of agriculture, population concentration grew exponentially, leading to the creation of villages and then towns and cities.

Agriculture also meant the domestication of animals for food and work, and many of our infectious diseases originated in animals; these are what we call zoonotic illnesses. The importance of this cross-connection between humans and animals has resulted in the One Health movement, which emphasizes that only by understanding the health of both humans and animals can we prevent diseases in our own species.

I was among the first supporters of One Health because it

addresses such a critical reason for today's increased risk of infectious diseases in humans.

Many infectious diseases, including poliovirus and variola major (smallpox), have adapted themselves solely to humans, with other variations (such as cowpox and monkeypox) affecting humans and other species. Zaire Ebola, the strain that caused the 2013–15 West African epidemic, is efficiently deadly to humans, with anywhere from one-third to one-half of the victims dying. Reston Ebola, the strain that played the lead in *The Hot Zone,* Richard Preston's 1994 bestseller, was fatal to primates but left humans virtually untouched.

Each infectious disease needs a certain human or animal population to sustain itself. Measles, for instance, one of the more effectively transmitted infectious diseases there is, likely requires a contiguous population of several hundred thousand; otherwise, it dies out.

Some biological agents can simply lie in wait for the right time to strike. If we have chicken pox as children, the varicella-zoster virus that caused it can remain latent in the body for decades. Then, when we're older and our immune system is weaker, it can break out in a form called herpes zoster, causing painful shingles. The bacterium *Bacillus anthracis* can lie dormant in spore form almost indefinitely, until it is inhaled or ingested or comes in contact with an open wound, at which point it will be reactivated and cause deadly anthrax in its unwitting host.

Once a disease effectively jumps from animal reservoir to human, it represents a new risk to its potential victim population, which has no biologic memory, and it takes time and trauma for that population (the surviving part of it) to form immunity. As civilizations grew and progressed, so did the speed and impact of infectious diseases. *Yersinia pestis* bacteria—the Black Death bubonic and pneumonic plague that wiped out between a quarter and a third of the European population in the fourteenth century—took only a decade to spread across Europe and continued to be deadly for more than a century.

But when the Europeans came to "settle" the New World two centuries later, they came upon peoples who were immunologically naïve to their bugs. The smallpox virus they brought with them cut the number of Timucuan Indians in Florida by half in six years—from roughly 722,000 in 1519 to 361,000 by 1524. Four years later, measles pandemic halved the population yet again. Similar courses had similar effects on other Native American civilizations, which the Spanish conquistadors took to mean that God favored their conquest and lust for gold.

As fast steamers replaced sailing ships and then trains superseded horse-drawn wagons, the efficiency of infectious disease spread picked up. That's pretty much where we were at the beginning of the twentieth century.

Statistically, the worst pandemic of the modern era occurred in 1918, when the so-called Spanish flu swept the globe. In reality, it was not Spanish at all. It was just that Spain, neutral in World War I and thus a country that did not censor its press, reported it honestly and therefore was erroneously stuck with the rap. Conservative estimates have traditionally pegged the worldwide death toll at 40 to 50 million, but recent analyses suggest it might have been twice that, dwarfing the toll of the brutal and bloody world war that immediately preceded it.

For reasons we will discuss, the 1918 flu was an influenza strain like none other in recorded history. Could something like this happen again? You bet it could. In fact, you bet your life it could. But with all of the advances in medical science and communications in the past hundred years, would we be better prepared to deal with it?

Don't be too sure.

The world is a far different place today than it was a century ago. In fact, the world is a far different place today than it was twenty-five years ago. And almost all of the changes that have taken place *favor* the microbe side of the war rather than the human side.

First, by its very nature, public health requires cooperation; communities and countries must come together. The worldwide smallpox eradication program worked because the two superpowers of the time—the United States and the Soviet Union—both agreed it was the right thing to do. Had either side not supported this effort, smallpox eradication would not have happened. And when those two gave marching orders, every other country in the world lined up behind them and saluted.

Since the fall of the Soviet Union, the world has changed. The nonprofit Fund for Peace's Fragile States Index was much higher in 2016 than a similar study would have showed in 1975, and it is more difficult now than it was forty years ago to get the international community to work together to achieve a common goal. Today, there are more than forty countries with no more than a marginal ability to govern.

It's not just Africa we're talking about. In the Americas, as of this writing, Venezuela and Colombia are on the verge of economic and political collapse, based on falling oil prices. Brazil's president has been impeached, its government is breaking down, and the state of Rio de Janeiro has declared a "public calamity." Puerto Rico, part of the United States, is virtually bankrupt. All of these disruptions in governance can lead to major catastrophes in public health.

Internal and external terrorism is a constant threat, and suspicion is persistent. As of this writing, a number of polio vaccination workers have been murdered in several areas of Pakistan, where hard-line Islamists oppose the campaign as contrary to the will of God and a secret attempt to sterilize the population.

Second, population tends to expand exponentially, and more and more humans and animals are concentrated in close proximity. We've already noted the human population explosion: In 1900, there were an estimated 800 million people on the earth; by 1960 that number had risen to 3 billion. Today it is around 7.6 billion. The World Health Organization (WHO) estimates the world population will reach 10 billion by 2050. Most of that growth will be in the megacities of the developing world, where the unsanitary conditions, including the lack of safe water and sewers, make Dickens's descriptions seem not so bad.

The concern we most often hear or read about animal populations around the world today is their serious loss in numbers, including the extinction of an increasing number of species. Yet there has also been an explosion in the population of food-production animals to feed the growing human population.

For example, in 1960, there were an estimated 3 billion chickens in the world; today there are approximately 20 billion. And since a chicken grows so quickly, the breast on your plate today may have been just an embryo as recently as thirty-five days ago. We can go through as many as eleven or twelve generations of chickens in one year.

Each one of these birds represents a potential test tube in which a new virus or bacteria can grow. And by the very nature of poultry production around the world, these birds are in close contact with humans; they share breathing space with those who care for them. The same is true for pigs. Today, there are more than 400 million pigs produced each year, and the pig happens to be the perfect genetic mixing bowl for the unstable and easily mutated avian and human influenza viruses.

To add fuel to the fire, it's expected that chicken and pig populations will increase by at least 25 to 30 percent over the next twenty years to help feed the rapidly growing human population.

Third, changes in global travel and trade have made us truly a one-world economy. People, animals, and goods are moving around the planet in greater numbers than ever before and at unprecedented speed. Until the last century, most of the world, particularly the developing world, was rural and isolated. Most people never traveled more than a few miles from the villages in which they were born. In 1850, it took almost a year to circumnavigate the globe on a fast sailing vessel. Today, we can go round the world in less than forty hours by airplane. The first scheduled commercial air flight occurred in 1914, transporting passengers across Tampa Bay. A hundred years later, 8 million people take commercial flights each day; that's more than 3.1 billion fliers annually.

The significance of the fact that any person can end up anywhere else on earth in a matter of hours is obvious. But just as significant is the idea that because of the global supply chain and the just-in-time-delivery practices affecting nearly all products and components, the impact of a pandemic will be far greater than one of similar virulence would have been in the past. As just one example, we may have among the world's best medical infrastructures in the United States, but virtually all of our generic lifesaving pharmaceuticals are manufactured overseas. Let's say there is a major epidemic in a region of India where many of our drugs come from. Lives will be lost in our own major cities because the critical medications just won't be available.

For the year ending June 30, 2014, air carriers transported 186 million passengers between the United States and the rest of the world. The same carriers also transported 9.54 million tons of freight between these countries. Globally, more than 150 million tons of freight were transported by planes. Every day, up to 60,000 large cargo ships are traversing the oceans of the world, bringing cargo containers from one continent to another, and with them a number of infectious disease vectors such as virus-infected mosquitoes and contaminated agricultural products.

Ironically, the ways we have organized the modern world for efficiency, for economic development, and for enhanced lifestyle—the largely successful attempts to transform the planet into a global village—have made us more susceptible to the effects of infectious disease than we were in 1918. And the more sophisticated, complex, and technologically integrated the world becomes, the more vulnerable we will be to one disastrous element devastating the entire system.

The fourth factor in our war on microbes is global climate change. Frankly, we don't know what the effects will be, but you can bet there will significant ones. Will malaria, which already kills between half a million and a million people each year, spread into areas farther from the equatorial region? This might happen with any of the tropical diseases, particularly those transmitted by mosquitoes, such as Zika. Will the midwestern winters no longer be cold enough to kill off the disease-causing agents of summer?

Malaria also highlights another important concept in public health: the distinction between epidemic and endemic diseases we alluded to earlier. Those more than half million fatalities in Africa far exceed any reasonable estimate of what Ebola could have caused in the 2014 outbreak. But malaria and other endemic diseases such as tuberculosis do not cause widespread panic in other countries or bring down governments. They don't lead to threats to shut down airports and close borders.

In contrast to a chronic condition, an outbreak, particularly one caused by a virus transmitted simply by breathing the same air as those already afflicted or suffering a mosquito bite we don't even feel or notice, creates a sense of panic combined with the struggle to understand the science and control the situation. This naturally leads to disproportionate disruption and impact. In the wake of the 9/11 attacks, a small amount of powdered anthrax sent through the US Postal Service to Capitol Hill and media figures, causing only twenty-two cases, still took billions of dollars to fix, closed down the Hart Senate Office Building across the street from the US Capitol for months, and paralyzed the mail delivery in the area. And anthrax is not a communicable disease the way Ebola and smallpox are. You don't catch it from another infected person.

Therefore, as serious as epidemics and pandemics can be in medical terms, we have to understand that certain kinds of deadly outbreaks can bring panic and disorder far in excess of their simple numerical effect—the frequent disconnect between what has the greatest potential to kill us or hurt us and what frightens us or just makes us uncomfortable.

A pandemic can shut down regional, national, or even international commerce, which in turn can lead to economic chaos, which in turn can lead to destruction of confidence in unstable governments. If a government's authority is shaky to begin with, the stress of a pandemic can lead to a failed state, which in turn can lead to anarchism and terrorism. At the same time, while the pandemic is occurring, other endemic and noninfectious diseases are still affecting the population, the combination of which can eventually tax or even break the existing healthcare delivery system.

In the three West African nations affected by the 2014 Ebola outbreak, crops were not harvested, schools shut down, borders closed, and the Peace Corps removed 340 volunteers. Because they were unable to receive medical care during the outbreak, almost as many people died from HIV, tuberculosis, and malaria infections as died from Ebola.

The very enemy we have devoted so much money and human resources to defeating since 9/11 can easily fill the leadership vacuum created by pandemic disease. In a very real sense, fighting infectious disease is, among other things, a matter of national security.

CHAPTER 7

Means of Transmission: Bats, Bugs, Lungs, and Penises

> *Nature being capricious and taking pleasure in creating and producing a continuous succession of lives and forms because she knows that they serve to increase her terrestrial substance, is more ready and swift in her creating than time is in destroying, and therefore she has ordained that many animals shall serve as food one for the other; and as this does not satisfy her desire she sends forth frequently certain noisome and pestilential vapours and continual plagues upon the vast accumulations and herds of animals and especially upon human beings who increase very rapidly because other animals do not feed upon them.*
>
> — LEONARDO DA VINCI

To move from where it is to the next available host, a microbe has to have a way of getting there. This is what we call means of transmission. Over the millennia, various pathogens have evolved different means of transmission, which are a prime factor in how much we need to worry about them.

The four categories enumerated in the chapter title do not represent a complete list, but rather the principal concepts we need to understand regarding disease spread.

Bats are a type of disease reservoir, which means a place where pathogens maintain themselves. For example, we believe, but have not yet definitely proved, that Marburg filovirus—a close cousin of Ebola—resides in fruit bats that live in locations such as Kitum Cave in Kenya's Mount Elgon National Park. The virus is excreted in the bats' guano and migrates from there. It's important to note that reservoirs need not be animals, or even alive. A reservoir can be a plant, a body of water, or any other host in which the pathogen can multiply and survive while it waits for its next spread. As we have seen with Marburg and Ebola, trying to discover or figure out the reservoir can be one of the great whodunit factors for a disease detective.

The mosquito is what is known as a vector: an arthropod that carries and transmits a pathogen into another host. Mosquitoes are the kings of vectors, our ultimate foe. In addition to prevention of illness through vaccines or other antibiotics, vector control is crucial in halting the spread of disease through mosquitoes and other insects. We will deal with this in depth in chapter 14.

Back in the 1400s, when mosquitoes accompanied mariners to the New World and on other voyages that took months or years, the mosquitoes aboard ships would die out before they could infect immunologically naïve populations. It took humans to do that. Today, a rat or mouse would most likely be noticed aboard a commercial airliner and dealt with before passengers embarked. But a mosquito can hitch a ride anywhere with virtual invisibility.

Lungs, which we all need to survive, are the scariest method of transmission, because through this method we can get sick simply from breathing—specifically, breathing in someone else's contaminated air. The 1918 influenza outbreak, which we have noted was the deadliest pandemic of the modern era, was an airborne transmission, as are all influenza strains. So-called respiratory-transmitted infections are the most likely candidates for quick spread because all they need is for their hosts to breathe.

Then there is the entire category of sexually transmitted infections, in which bodily fluids are exchanged between sexual partners. This has always been a touchy subject for public health authorities because people don't like to talk about it and it is difficult to obtain honest reporting and good statistics. Despite the fact that we are all here as the result of a sex act, meaningful discussion remains one of the great societal taboos. With sexually transmitted infections, epidemiology must venture far into the realm of sociology. When it arrives there, what we tend to discover—or relearn—is how difficult it is to get people to alter their habits, and that, in too many instances, women are denied agency over their own sexual destiny.

Syphilis, an age-old scourge caused by the *Treponema pallidum* bacteria, is one of those ailments no group wanted to claim and all groups wanted to blame on some other. After an invasion by the French in the late 1400s, Neapolitans labeled it "the French disease." The French, on the other hand, called it "the disease of Naples." The Russians called it "the Polish disease," while the Polish and Persians called it "the Turkish disease." The Turks called it "the Christian disease," the Tahitians called it "the British disease," the Indians called it "the Portuguese disease," the Japanese called it "the Chinese pox," and so on and so on. A similar worldwide paranoia greeted the advent of HIV/AIDS, which is one of the reasons Jim Curran at the CDC was so insistent that the world scientific community quickly adopt a name that would be completely neutral and the same in every language.

While many of us came of age during the so-called sexual revolution of the 1960s and succeeding decades, we should remember that there was only a narrow window of history in which sex couldn't kill you: between the widespread availability of sulfa drugs and antibiotics in the 1940s that combatted the bacterial STDs and the coming of AIDS in the early 1980s. And yes, we do now have drug cocktails that keep the levels of HIV under control, but AIDS is still a worldwide killer in much of the poor and developing world, where populations don't have access to modern medicines. And lest we get too complacent about syphilis, gonorrhea, and the other sexually transmitted pathogens, as we will see later in this book, the future effectiveness of antibiotics is a very shaky proposition. All of this is to say that our common enemy never gives up the fight.

Another aspect of disease transmission by penis that we cannot ignore is rape as a weapon of war. All decent people are appalled by crimes of sexual assault, and horrified when one of them results in a sexually transmitted disease. But throughout history, rape has also been used as a means to terrorize and help conquer an enemy's civilian population, and today we are seeing it employed on a strategic basis in conflicts in Africa and the Middle East. Suffice it to say that every rapist is a craven and unredeemable collaborator with humankind's common enemy and guilty of the most damning charge it is possible to bring against a human being: a crime against humanity.

A complex web of factors determines which pathogens can kill us, hurt us, or merely inconvenience us. At the heart of this web is a single critical consideration: How does the microbe get transmitted? In our disease-control business, transmission is defined as any mechanism by which the microbe is spread through the environment or to another human or animal. These mechanisms may include direct body contact with a human or animal; breathing air that was just exhaled from another person or animal, an aerosol purposefully sprayed into the air, or a mist from a nearby building's cooling tower; consuming food or water; physical contact with a surface such as a door handle; a mosquito or tick bite; a blood transfusion; or contact with blood on a previously used or contaminated needle.

While all of these mechanisms are significant spreaders of specific diseases, the ability to transmit a microbe by merely breathing it into our lungs is the most dangerous. We call this airborne transmission. In the real estate business, they say, it's "location, location, location." In public health, it's "airborne, airborne, airborne."

The potential for airborne transmission of a virus was demonstrated in stark clarity with a measles outbreak investigation I led in Minnesota in 1991. The outbreak occurred in conjunction with the Special Olympics and an infected twelve-year-old male track-and-field athlete from Argentina. He was in the highly infectious early stage of illness when he stood for several hours near home plate during the opening night ceremony in the enclosed Hubert H. Humphrey Metrodome. Other competitors, game officials, and support staff came down with measles after exposure to the young athlete. Two of the subsequent cases were Minnesota residents who did not know each other and did not attend any other Special Olympics event beyond opening night. But both sat in the same upper-deck section, more than 400 feet from home plate. Stadium airflow circulation data for that night supported the conclusion that air from where the athlete entered the stadium or from where he stood at home plate would have been pushed toward the two spectators who developed measles.

The most notorious of these airborne-transmitted diseases is influenza, and while we classify flu strains by subgroups of two of its surface proteins—hemagglutinin and neuraminidase, HA and NA—for our purposes here we'll divide flu viruses into two groups. The first is seasonal flu: the kind that makes you feel miserable, fills hospitals most winters, causes widespread absenteeism from schools and workplaces, and kills between 3,000 and 49,000 people each year in the United States. The other is pandemic influenza, which occurs when a new flu virus emerges out of the animal world through mutation or reassortment so that it can infect and be transmitted by humans. Generally, seasonal flu is a remnant of a strain of the flu virus that once caused a pandemic.

Throughout history, influenza and its ability to quickly kill many millions during a global pandemic has earned it the status of king of infectious diseases. An infected human can efficiently transmit the flu virus to people around him, and unlike someone infected with Ebola, say, he can do this even before he shows signs of being sick. All that is required is to breathe the contaminated air just exhaled or coughed from the lungs of an infected person. Imagine that person on a plane or subway car or at a shopping mall or sporting event where we all share one big common bucket of air. And remember how many people fly around the world each day as we consider how fast a disease like influenza may spread across the globe. Unfortunately, I am certain we are actually more vulnerable worldwide to an influenza pandemic today than we have been at any time over the past five centuries.

Airborne transmission is also a major concern with regard to the use of microbes for terrorism attacks. We now know that highly infectious *Bacillus anthracis* spores, the cause of anthrax, can travel many miles in the air when released in a simple, relatively easy-to-prepare powder from an airplane routinely used in agricultural crop spraying or mosquito control. Breathing in just a few of these spores is enough to cause a life-threatening reaction.

The next most concerning category of disease transmission is really a toss-up. As long as AIDS cases continue to increase in numbers around the world each year as a result of direct-contact spread, namely, through sex or through birth to an HIV-infected mother who is not receiving appropriate HIV drug treatment, this mode of disease spread is of critical public health importance. I don't include HIV transmission from sharing contaminated needles here, as it is technically classified as indirect transmission. It, too, remains an important part of the HIV risk picture, but direct-contact spread is still the most critical aspect of HIV today. Yet while this disease remains a high public health priority because of its international morbidity and mortality, particularly in Central Africa, the development and availability of the drugs that have made it a "livable" chronic condition have taken away its emergency or crisis profile in wealthier countries.

The second transmission category in the toss-up is vector-borne diseases—those transmitted by mosquitoes, ticks, and flies. We have now moved many species of mosquitoes that can transmit any number of infectious diseases to humans and animals around the world inside aircraft and cargo ships. When mosquitoes originally native only to Southeast Asia are transported to the Americas inside tires in the holds of cargo ships, they proliferate quickly in their new homeland. Never before in the history of humankind has there been the current extensive number of species of microbe-carrying mosquitoes on every continent except Antarctica. As a result, in just the past fifteen years we have witnessed the major global spread of diseases like dengue fever, West Nile virus, chikungunya, and Zika. And we still have to consider the reemergence of yellow fever and highly drug-resistant malaria. This disease transmission category also does not bode well for us as it relates to global climate change. A warmer world presents us with the potential for less overall precipitation in some regions. But when it does rain, it will be in monsoon-level amounts. This means that disease-causing mosquitoes will be sharing even more territory with large human populations.

The final transmission category we are calling "current world conditions": an amalgam of factors within three very different, yet highly microbe-rich environments. First is the exploding human population in the megacities of the developing world and the packed, horrible conditions in which the unfortunate residents live. Second is the human contact with animals in the rain forests of Asia, South America, and Africa, the ultimate fertile grounds for new and dangerous human pathogens, which are now spilling out into the inhabited world. Third is the high-intensity animal-production facilities around the globe that represent millions of new, living, animal "test tubes" for microbes, born each day.

Why were we surprised that Ebola virus, a disease that to date is spread by direct contact with contaminated body fluids, moved as quickly and efficiently as it did in the villages and slums of the three impacted West African countries? Why are we surprised by the unprecedented increases in avian influenza viruses—the precursors for a human pandemic influenza strain—associated with exploding global poultry production? Why were we surprised by the rapid spread of Zika virus throughout the Americas when the *Aedes aegypti* mosquito, the vector for this disease, is now widespread within this area?

If there is a lesson here, it is that we have to think seriously about these things. And we haven't been.

CHAPTER 8

Vaccines: The Sharpest Arrow in Our Quiver

The return on investment in global health is tremendous, and the biggest bang for the buck comes from vaccines. Vaccines are among the most successful and cost-effective health investments in history.

— Seth Berkley, MD

It's hard to overstate the impact of vaccines on our history and on our lives.

The term "vaccine" hearkens back to the work of Edward Jenner, who referred to cowpox, the disease to which he exposed patients to immunize them from smallpox, as *Variolae vaccinae,* Latin for "smallpox of the cow." With the success and popularity of this means of inoculation against one of history's great killers, all such methods became known as vaccination.

But while we rightly consider Jenner the father of vaccination, the basic concept probably goes back a thousand years. Recognizing that scratching or cutting the skin and inserting a small amount of smallpox pus could confer immunity, Chinese healers in the tenth century employed a practice known as variolation. An alternate method was to let the pus dry into a powder and then blow it up the nose. Though these practices did keep many recipients from getting full-blown smallpox, they were not without significant risk; they could cause the disease—sometimes fatally—as well as transmit other dangerous microorganisms, including syphilis bacteria, into the scratched or cut skin or through inhalation into the lungs. But as the best means of inoculation available until Jenner's time, they were adopted by many cultures.

Jenner's inoculation method changed everything and ushered in the modern era of vaccines. The benefits were recognized at different times in different nations. In some, skeptics physically attacked vaccinators as charlatans or worse.

In 1777, General George Washington mandated smallpox inoculation for every member of the Continental Army. In 1806, with Jenner's method in wide use, President Thomas Jefferson publicly endorsed vaccination. "Medecine [*sic*] has never before produced any single improvement of such utility," he declared. Seven years later, the US Vaccine Agency was established under President James Madison, who instructed the US Post Office to carry smallpox vaccine without charge. In 1885 Louis Pasteur announced his vaccine for rabies, a disease that had carried a 100 percent fatality rate. Jefferson's observation was now difficult to deny.

So compelling was the case for the early vaccines that in 1905, the Supreme Court ruled in *Jacobson v Massachusetts* that the benefit of compulsory smallpox vaccination to public health took precedence over an individual's personal agency to refuse.

From around that time, scientific discoveries in infectious disease etiology, antitoxins, and means of transmission ushered in the great age of vaccine innovation. One glance at the CDC tables comparing annual morbidity and mortality in the United States in the twentieth century and in 2014 from a group of common infectious diseases presents a striking picture.

The annual number of twentieth-century US pertussis—whooping cough—cases averaged 200,752 before vaccine was available. In 2014 the number was 32,971: an 84 percent decrease. By the same time measures, measles went from an average of 530,217 cases per year to 668 by 2014: a 99 percent decrease from before vaccine was used in children. In 1964, in the last great US outbreak of rubella, a disease that can be devastating to the unborn children of affected pregnant women, 2,100 babies died and another 20,000 were born with severe, lifelong disabilities. Today cases of mumps and rubella have decreased by a similar 99 percent. Tetanus, with an extremely high mortality rate, went down 96 percent. And polio, diphtheria, and smallpox all went to zero cases.

At the turn of the twentieth century, the US infant mortality rate—the rate of deaths among children in the first year of life—was 20 percent, in some cities as high as 30 percent. Of those fortunate 70 to 80 percent who survived, another 20 percent died before reaching their fifth birthday. Later in the century, similar deaths in children were greatly reduced due to vaccination and improvements in basic sanitation.

From 1900 to 1904, an average of 48,164 cases and 1,528 deaths caused by smallpox were reported each year in the United States. Outbreaks occurred periodically after 1905 and then ended in 1929. Sporadic cases continued until 1949. The absence of smallpox cases in the United States for the past sixty-seven years is one of the most remarkable public health achievements of all time, considering the death, disfigurement, and suffering the virus caused for centuries.

In 1954, Jonas Salk, a virologist at the University of Pittsburgh School of Medicine and developer of the first polio vaccine, became an international hero to the generations of parents who worried every summer when their children went to a playground, swimming pool, or movie theater—anywhere people congregated and the poliovirus silently lurked. They were haunted by images of row after row of iron-lung respirators and boys and girls in leg braces and wheelchairs. Now there was a prospect of those images disappearing from the modern world.

On April 12, 1955, in what became one of the most famous quotes of the decade, legendary broadcast journalist Edward R. Murrow asked Salk on the live CBS program *See It Now,* "Who owns the patent on this vaccine?"

With matter-of-fact modesty and a shy smile, Salk replied, "Well, the people, I would say. There is no patent. Could you patent the sun?"

That was it—the apotheosis from man to immortal. Jonas Salk was every parent's selfless deliverer from fear.

Salk's archrival, Dr. Albert Sabin of the Cincinnati Children's Hospital Medical Center, later developed a vaccine based on live attenuated virus—a virus that has been changed so it doesn't cause disease but still grows in humans or animals—which could be administered on a sugar cube rather than injected into the arm. Both vaccines were highly effective in their mutual objective of protecting humans from polio.

Even without a patent, the vaccines were economically viable, which encouraged a number of companies to get into the polio vaccine business and reaffirmed Jefferson's observation that vaccines were there for the good of all.

And that, in turn, created a vital and continuous manufacturing demand. The vaccine *business* flourished. Five major pharmaceutical companies produced Salk's vaccine. Between 1955 and 1962, 400 million doses were administered in the United States alone. Just about everyone was inoculated against smallpox and polio.

Over the course of the 1960s and 1970s, children in the United States and other developed world countries began receiving a standard lineup of immunizations before they started school. These included diphtheria, tetanus, and pertussis (DTP), and later measles, mumps, and rubella (MMR) and chicken pox. Most school districts required proof of immunization before parents could enroll their kids. Vaccination against deadly rabies was standard procedure for anyone bitten by a suspect animal that couldn't be caught and examined or that was captured and found to have rabies. Newly recruited soldiers and sailors lined up for vaccinations against anything the military feared they might face, including yearly influenza shots. There was an ongoing need for vaccines, and pharmaceutical companies were eager to participate in a lucrative business model that supported public health on a mass scale.

These staggering advances are all due to vaccines. It is no exaggeration to say that vaccine remains, along with basic sanitation, the sharpest and most effective arrow in our public health quiver. How we aim that arrow will determine our future.

So successful was the effort to curtail or eradicate the range of childhood diseases that the public started taking their absence for granted. This, among other things, has given rise to an anti-vaccine movement, whose members are wary of vaccines, particularly childhood vaccines, believing that they may cause autism, or even the diseases they are supposed to prevent. There is no scientific evidence to support these charges, but that doesn't stop a good many sophisticated, educated people from backing away from vaccines that were once considered miraculous. Ironically, this resistance recalls the dawn of vaccines, when smallpox vaccinators were harassed and attacked by suspicious opponents. But they, at least, had the excuse of lack of established knowledge.

Today's opponents have no such defense. Measles, for example, which is usually self-limiting but can become very serious in some individuals (in the immunocompromised, the fatality rate can be as high as 30 percent), was eliminated from the United States by the year 2000. But it has returned, caused by infected children from other countries that still have measles traveling to the United States and exposing our unvaccinated children. And that transmission can happen easily, as when an infected guest visited Disneyland in California in 2015. The outbreak sickened 147 people in the United States, including 131 in California. Whether this was due to the complacent belief that the disease was a thing of the past or misplaced fear of the highly effective vaccine doesn't matter. The result was needless sickness—some of it quite severe—widespread fear, and economic costs.

It isn't just complacency and the antivaccine crowd that challenge vaccine development. The basic economics have changed.

Today, the pharmaceutical business model for routine immunizations and travel-related protection, such as against yellow fever and typhoid, still holds, even though fewer manufacturers remain in the business and bulk purchasers such as the government and insurance companies have brought prices and profit margins way down on certain vaccines. In 2002, Wyeth pharmaceutical company stopped producing DTP and flu vaccines. The move had negligible effect on company profits but created rationing for both vaccines in the following year.

But now we have new and different vaccine needs and the business model has become more complicated. Pharmaceutical manufacturers are noting that vaccine production is no longer where the major action is. In 2014, the worldwide pharmaceutical industry was estimated to have more than $1 trillion in annual revenues. Just the five leading drugs around the world generated more than $49 billion in sales. This included three autoimmune drugs—Humira ($12.54 billion), Remicade ($9.24 billion), and Enbrel ($8.54 billion); Solvaldi, for hepatitis C ($10.28 billion); and Lantus, a drug for diabetes ($8.54 billion). Overall, the ten biggest-selling pharmaceutical products in 2014 generated combined sales of $83 billion.

In contrast, in 2014, the top five vaccine manufacturers in the world had combined sales of $23.4 billion, a mere 2 to 3 percent of the trillion-dollar drug market.

Let's get one thing straight about vaccines: It's not like in the disease outbreak thriller novels and movies. A bunch of scientists in a lab don't suddenly find the magic formula, bottle it up, and have a medical flying squad race to the scene and inject it into the arms of the stricken, who, miraculously, recover in a matter of seconds or minutes. For one thing, vaccines are almost always for prevention rather than treatment. For another, once you've got the proof-of-concept "formula" that appears to work in the lab and then in animal models, you've got a long way to go before you can even submit the vaccine for FDA approval and then create and ramp up production facilities, not to mention figuring out how to pay for all of this.

Vaccines are not like other kinds of drugs, and comparatively speaking, they are hard to make. The production of the Lipitor you take for your cholesterol, the Metformin you take for diabetes, the Prozac you take for depression, or the Viagra you take for erectile dysfunction—all maintenance drugs of one kind or another—can be likened to building a Chevrolet on a General Motors assembly line. Production of a vaccine, on the other hand—particularly a new vaccine—is more like growing lettuce in a field in California. By the time the Chevy gets to your garage or the lettuce gets to your table, each is going to be pretty much what you expected. But the process for manufacturing the car is a lot more predictable, repeatable, and scalable than the process for growing lettuce, which is subject to weather, ground conditions, drought or flood, insects, and any plant-based diseases that happen to be circulating in the area.

What we're talking about is the difference between a chemical agent and an essentially biologic agent; that is, chemical synthesis versus biological growth. For decades our vaccines have grown in cell cultures, in eggs, or on the skin of animals such as calves. This is a time-consuming process with a number of difficult-to-control manufacturing variables. And most of the vaccine production for influenza requires a whole lot of chickens laying a whole lot of eggs. The more modern cell culture technology, in which a seed virus is introduced into an existing cell line and grown in capacity in a fermenter, is faster and more efficient, but it's still a biologic process.

Just as a vaccine is different from a maintenance drug in terms of manufacturing and makeup, it is fundamentally different from an economic perspective. A pharmaceutical company can count on a regular and predictable market for the maintenance drugs its customers will take every day, often for the rest of their lives. For the major noncommunicable illnesses like cancer, manufacturers know they will have a steady market because the diseases are not going away anytime soon, and they can charge a lot of money for their drugs as long as their patent monopoly lasts.

In contrast, the need for a particular vaccine is unsteady and unpredictable. By the time you need one that is already licensed, it is often too late to ramp up production. During the 2009–10 H1N1 flu pandemic, the number of cases of the second, critical wave peaked in the United States in October 2009. The number of doses of vaccine shipped peaked at the end of January 2010, by which time the number of cases had dropped sixfold. Even then, the number of doses shipped in the United States was less than 125 million. This was far short of the number needed to vaccinate every American, particularly given that children required two doses.

To be dispensed in the United States, vaccines have to go through the same sorts of FDA-mandated clinical trials as other pharmaceuticals. As vaccine development progresses, there are various internal tests, and then animal testing. Then there are three phases of human trials. Phase I tests safety. Phase II tests various dosage levels for safety and effectiveness. Phase III tests the actual effectiveness of the drug or vaccine on a large enough cohort of human subjects to allow for variations in response, factoring in considerations such as how the vaccine affects children, teens, persons over sixty-five, persons with an immunocompromising condition, pregnant women, and so on.

Generally, Phase III trials are double-blind, meaning that neither the subject nor the administrator knows which subjects are given the actual drug and which are given a placebo. At the end of the trial, that information is revealed and the outcomes are compared. Sometimes the trials are stopped early when an independent monitoring board determines that during the trial, the vaccine has clearly demonstrated it is or is not working, or there are patient safety issues emerging. Phase III trials can get extremely expensive, and pharmaceutical companies don't like to undertake them unless they think they have a pretty strong prospect of obtaining FDA approval. Today a pharmaceutical company can expect that getting a new vaccine licensed will take more than a decade of work and a billion dollars of investment.

Pharmaceutical executives know that the process from the beginning of Phase III through its results, submission to the FDA's Office of Vaccines Research and Review, and that office's complete review and evaluation literally can take years. We call this Phase III evaluation "the valley of death": the point at which substantial research, development, testing, and licensing costs are piling up but no revenue is being generated.

To understand this phenomenon, let's back up a couple of steps. Vaccine development often starts with grants and contracts from the NIH and science- and health-oriented foundations, as well as "angel" investors. Much of the research originates from the academic sphere. This initial development step, if successful, can get the vaccine to the prototype stage and through Phase II trials. But then the product enters the valley of death. Now the prospect of huge expenses comes into focus and the researcher-developer has to make some fundamental decisions.

What are the chances the vaccine will make it through Phase III trials and prove itself effective and without serious side effects? What are the chances the vaccine will find a large and steady market if it does successfully make it through Phase III trials and wins FDA approval? How much will manufacturing facilities cost? What about the added time and expense of having to go through other countries' regulatory procedures? How do you decide to allocate research and development dollars, including those for Phase III trials for diseases that might best be considered "potential global calamities waiting to happen" but might not reveal themselves for years or even decades to come? The West African experience with Ebola and the Americas' experience with Zika virus infections are two examples of this challenge.

This makes sense. Corporations cannot ignore economic realities. They have to demonstrate to their boards that they are acting rationally from a business perspective. While we all applaud corporate social responsibility, we cannot expect it to be a business model. As Dr. Rajeev Venkayya, president of the Global Vaccine Business Unit of Takeda Pharmaceutical Company and former director of Global Health Delivery for the Bill & Melinda Gates Foundation, told a conference at the National Academy of Medicine, "Pharmaceutical companies want to do the right thing, but they don't like risk or tolerate it well."

Philanthropic funding still plays a role in the research and development of vaccines and their subsequent purchase, as was modeled by the March of Dimes and the polio campaign. The Bill & Melinda Gates Foundation is partnering with academic research groups, pharmaceutical companies, and product development partnerships to try to develop a vaccine for HIV/AIDS and a more effective one for malaria, two of the biggest infectious killers in Africa. And there are other examples.

But as Bill Gates said to us when Mark and I met with him in his Seattle-area office, "People invest in high-probability scenarios: the markets that are there. And these low-probability things that maybe you should buy an insurance policy for by investing in capacity up front, don't get done. Society allocates resources primarily in this capitalistic way. The irony is that there's really no reward for being the one who anticipates the challenge."

Every time there is a new, serious viral outbreak, such as Ebola in 2012 and Zika in 2016, there is a public outcry, a demand to know why a vaccine wasn't available to combat this latest threat. Next a public health official predicts a vaccine will be available in *x* number of months. These predictions almost always turn out to be wrong. And even if they're right, there are problems in getting the vaccine production scaled up to meet the size and location of the threat, or the virus has receded to where it came from and there is no longer a demand for prevention or treatment. Here is Bill Gates again:

> Unfortunately, the message from the private sector has been quite negative, like H1N1 [the 2009 epidemic influenza strain]: A lot of vaccine was procured because people thought it would spread. Then, after it was all over, they sort of persecuted the WHO people and claimed GSK [GlaxoSmithKline] sold this stuff and they should have known the thing would end and it was a waste of money. That was bad. Even with
>
> Ebola, these guys—Merck, GSK, and J & J [Johnson & Johnson]—all spent a bunch of money and it's not clear they won't have wasted their money. They're not break-even at this stage for the things they went and did, even though at the time everyone was saying, "Of course you'll get paid. Just go and do all this stuff." So it does attenuate the responsiveness.

This model will never work or serve our worldwide needs. Yet if we don't change the model, the outcome will not change, either.

Let's consider one example. Each year, starting around September, we're all admonished to get our influenza vaccinations. And yet, each year, we all hear from someone, "I got the vaccine last time and I still got the flu!" A couple of years ago, it happened to me: Even though I got the shot, I still wound up in bed for a week with influenza.

The fact is, influenza vaccine is one of our least effective vaccines, and the only one that has to be changed every year. That is partly because influenza strains shift so easily that public health officials have to make an educated guess about which strain or strains will be predominant in a given year, and they have to do it many months in advance by observing what is going on in the other hemisphere of the world. We follow what is happening with influenza virus strains in the Southern Hemisphere when it is their fall (our spring) to predict which influenza viruses will likely be with us the next winter. Some years that educated guess is more accurate than others.

So is it worth getting the vaccination each year? I give that a qualified *yes.* It might or might not prevent you from getting flu. But even if it is only 30 to 60 percent effective, it sure beats zero protection.

What we really need is a game-changing influenza vaccine that will target the conserved—or unchanging—features of the influenza viruses that are more likely to cause human influenza pandemics and subsequently seasonal influenza in the following years.

How difficult would such a game-changing influenza vaccine be to achieve? The simple truth is that we don't know, because we've never gotten a prototype into, let alone through, the valley of death.

We need a new paradigm—a new business model that pairs public money with private pharmaceutical company partnerships and foundation support and guidance.

What might that look like?

Going back to our war analogy, when the Department of Defense decides it needs a new weapon system, it puts out general specifications and solicits bids, but it doesn't expect the large defense contractors to develop that weapon, test it, and then hope the government wants to buy it in quantities sufficient to make it profitable. Instead, bids are evaluated and a contractor or consortium of contractors is selected. If we're serious about having vaccines for a wide range of potentially destructive or antibiotic-resistant infectious diseases, we need to strongly consider the government's involvement—not just in initial research and development, but also in actually bringing the vaccine to market.

We'd like to see the paradigm shift throughout the world, but the United States, as is so often the case, will have to lead. We surely welcome the countries of the European Union, China, and even India to provide science and policy leadership as well as financial resources. But we can't wait for an international consensus; the infectious bugs are now gaining on us at breakneck speed. The US government must increase its support of the development of vaccines that will address our Crisis Agenda, and coordination among government, academia, and industry will be needed to ensure that the vaccines with clear potential make it across the valley of death.

The US government has tried to make a real difference in the critical vaccine arena. Foreign and terrorist threats reliably get official attention. Following 9/11 and during the subsequent anthrax attack, Health and Human Services secretary Tommy Thompson asked me to serve as a special adviser to him and the highly competent and seasoned team of bioterrorism and public health experts he had assembled. He had become aware of my experience in these areas of concern after reading my book *Living Terrors,* as well as through my numerous calls and meetings with his senior staff in the days following 9/11. I subsequently spent more than three years as a special adviser to the secretary's office on a part-time basis while still serving as the director of CIDRAP. To my pleasant surprise, I quickly learned that Secretary Thompson understood, as did few other senior government officials, the critical importance of public health preparedness.

One of the efforts I was involved with was called Project BioShield. It was the brainchild of Stewart Simonson—one of the secretary's closest advisers and the first assistant secretary for public health emergency preparedness—and Major General Philip K. Russell, MD—former head of the US Army Medical Research and Materiel Command and an expert on vaccine development. In addition, the late D. A. Henderson; Anthony (Tony) Fauci, MD, director of the NIH's National Institute of Allergy and Infectious Diseases (NIAID), who came up with the name; the late John LaMontagne, PhD, deputy director of NIAID; William Raub, PhD, former acting director of the NIH and then science adviser to Secretary Thompson; and Kerry Weems, a career executive at the Department of Health and Human Services (HHS), made up the team that brought BioShield to reality. As a result of their visionary and groundbreaking work, Congress appropriated $5.6 billion in fiscal year 2004 to the Project BioShield Special Reserve Fund to support the goal of acquiring chemical, biological, radiological, and nuclear (CBRN) medical countermeasures over a ten-year period. The hope was that having such a large, precommitted government fund would incentivize the pharmaceutical industry to invest its resources in multiyear countermeasure projects.

By guaranteeing the market, the fund attracted a number of smaller to midsize pharmaceutical companies to participate in countermeasure product development, including new vaccines. Unfortunately the $5.6 billion fund was not adequate to entice larger companies, which have unique expertise in vaccine production, to get involved in this work. Nonetheless, a number of countermeasure products, particularly related to terrorism response, were secured. This fund has run its ten-year course (2004–14) and exhausted the advance commitment support. It now requires an annual appropriation from Congress, which is always fraught with uncertainty and therefore is a deterrent to companies that understandably want to commit only to multiyear projects.

Throughout the often shaky relationship between government,

the public health establishment, and the pharmaceutical industry, you will continually hear laments about the severe difficulty of obtaining ongoing budgetary commitments on anything that cannot be labeled defense spending or Homeland Security spending. Defense funders are used to requests for multiyear budgets. You can't develop and build a weapon system in a year. But almost everything we do in public health and medical countermeasures also takes longer than one fiscal year or one funding cycle. When it comes to funding, the single most common aspirational word we hear is "sustainability."

In 2006, Congress established the Biomedical Advanced Research and Development Authority (BARDA). It is intended to provide an integrated, systematic approach to the development and purchase of the necessary vaccines, drugs, therapies, and diagnostic tools for public health medical emergencies. Project BioShield is now part of BARDA. Its annual appropriated budget must now cover the development of all CBRN measures. In 2016, the budget was approximately $1.8 billion, with no dedicated funds for emerging infectious diseases, including vaccines or drug treatments. And the need to go to Congress and ask for new money every year has all but killed the possibility of major long-term projects, such as the development of game-changing influenza vaccines.

While I respect the efforts of the BARDA staff, the way they have to do business is just not sufficient for what we need to obtain the vaccines for worldwide pandemics or epidemics of critical regional importance. Far too often BARDA is pressured by key members of Congress to prioritize the development and procurement of certain countermeasures when those countermeasures are made by companies in their districts or states. While such influence is not always obvious to the public, one need look no further than BARDA's decisions regarding the procurement of anthrax vaccine to realize the power of one company's lobbying efforts in Congress and in turn, BARDA. In addition, I believe that far too often, when BARDA senior staff members have been called before Congress to testify on the status of their programs, they have provided a "glass half-full perspective" when, in fact, the glass was damn near dry. This surely has been the case with pandemic influenza preparedness. The current federal government effort to secure needed new vaccines may not be a recipe for disaster, but it is certainly a recipe for getting little done in advance of a crisis, as recent history has proven.

Lately, others outside the US government have realized that increasing threats from emerging infections demand improved global preparedness. Three independent initiatives led by the WHO, the Norwegian Institute of Public Health, and the Foundation for Vaccine Research have come up with lists of "Priority Pathogens" for top funding. A pathogen's position on this list is based on its likelihood of occurrence, its potential impact on global health, and the reasonable chances of coming up with a safe and effective vaccine.

The Foundation for Vaccine Research proposed a global vaccine fund with an initial capitalization of $2 billion to take on the first of the forty-seven diseases with no vaccine or only partially effective ones. The fund would be intended to move vaccine prototypes for Crisis Agenda diseases such as MERS, Ebola, and Zika from the laboratory and through the valley of death so that they would be ready and available when outbreaks occur. Citing the fact that there are now only four major manufacturers that focus on vaccine development—GlaxoSmithKline, Merck, Pfizer, and Sanofi Pasteur—the authors called for seed money to come from governments, foundations, the pharmaceutical industry itself, and nontraditional but related sources, such as the insurance and travel industries. To justify the funding, they note that due to the lack of a proven Ebola vaccine, the 2013–15 crisis cost upward of $8 billion. However, there was no economic incentive for bringing an Ebola vaccine to market because the target population in Africa couldn't afford it.

Lawrence Summers, Charles W. Eliot Professor and president emeritus of Harvard University, as well as former secretary of the Treasury, was quick to tell us, "I would not dream of calling myself an expert in this field." That may be, but his analyses and perspectives on public health are consistently insightful. Delivering the keynote address for the release of the Global Health Risk Framework Commission report *The Neglected Dimension of Global Security: A Framework to Counter Infectious Disease Crises,* he said:

> With respect to vaccines in general and with respect to the capacity to develop vaccines as rapidly as possible in the wake of an emergency, it is essential that we invest more. This is the quintessential problem for which we cannot rely on the private sector. No one would permit, nor should anyone want, to profit immensely from having the scarce vaccine or antibody at the moment of pandemic. Therefore, the private sector will not be able to capture even a small fraction of the social benefit from a valuable preventative.

The Foundation for Vaccine Research, the WHO, and the Norwegian Institute efforts are highly commendable and a great first step. But who will pay for this major new international effort? How much will they pay, and for how long? Who will decide which vaccines are rushed to the head of the line for investment? Who will be responsible for the oversight of both the public and private sector partners? The list of questions goes on.

While hope is not a strategy, I am hopeful that there is a new and frankly exciting development in the vaccine world. As a result of the ongoing conversations among leaders involved in the three above-noted organizations, major foundations, the World Economic Forum, major vaccine manufacturers, and the US government, a new organization is emerging: the Coalition for Epidemic Preparedness Innovations (CEPI).

I have participated in two of CEPI's four working groups, and having seen it from the inside, I am optimistic that the vision described on the CEPI website for this yet emerging coalition is potentially game changing: "Epidemic outbreaks of infectious diseases will be managed at an early stage to prevent them becoming public health emergencies that result in loss of life, undermine social and economic development and emerge into humanitarian crises."

Taking an end-to-end approach from initial vaccine development to application, CEPI will focus on essential gaps in the process due to market failure. The initial focus will be to move new vaccines through the entire procedure, from preclinical to proof of principle in humans and to the creation of platforms that can be used for rapid vaccine development against unknown pathogens. How we will find the sustained funds to make this effort a reality is a huge unanswered question. Still, I believe this group does represent the best chance we have ever had for creating a sustainable international response for realizing a viable and dependable pipeline for critical vaccines, and we should all pay close attention to CEPI's progress. Our lives could one day depend on it.

CHAPTER 9

Malaria, AIDS, and TB: Lest We Forget

> *If you look at three diseases, the three major killers, HIV, tuberculosis and malaria, the only disease for which we have really good drugs is HIV. And it's very simple, because there's a market in the United States and Europe.*
>
> —Jim Yong Kim, MD, President, World Bank

In 2014, the most recent year for which statistics were available from the WHO, there were an estimated 36.9 million persons living with HIV worldwide and 1.2 million deaths from AIDS. There were an estimated 9.6 million cases of tuberculosis and 1.1 million deaths, according to 2015 statistics. There were 214 million cases of malaria and 438,000 deaths that same year. And yet that mass of human misery and mortality does not capture even a small fraction of the headlines and media attention as would ten cases of smallpox in a major city anywhere in the world.

We keep coming back to the reality that what kills us, what hurts us, and what scares us are not one and the same. For those of us in the so-called First World, these three major infectious killers are now comfortably assimilated into our threat matrices, along with other such everyday possibilities as automobile accidents and street crime. We know they exist; we just don't think about them much.

It wasn't always like that. Those of us who lived through the 1980s recall the terror evoked by AIDS, when a diagnosis of the newly discovered human immunodeficiency virus was a death sentence. In our grandparents' and great-grandparents' times, tuberculosis could mean a quick and painful death or a slow wasting away, with no treatment except rest and cool, dry air. Down through the centuries, malaria was a serious risk for everyone living in many parts of the world, including my home state of Minnesota.

Today, though we still have no cure or preventative, an effective cocktail of drugs keeps most of the deadly effects of HIV at bay. TB requires a long and rigorous regimen of antibiotics to cure, and malaria is rare in Westernized regions of the world.

While we have been relatively complacent, these three diseases remain major threats to world health, particularly in areas and countries too poor to afford treatment or adequate medical infrastructure. This book is primarily about "crisis" infectious agents: pathogens of pandemic potential and pathogens of critical regional importance. But this work would be incomplete and I would be remiss to neglect these three. And I never forget that there are other infectious diseases of major public health importance around the world, including hepatitis C, water- and foodborne diseases, bacterial pneumonia, other, neglected tropical diseases, and even human rabies, which kills upward of 50,000 people a year, primarily in Asia and following a bite from a rabid dog.

Fortunately, some people and organizations are trying to alter the status quo, putting a lot of resources into the effort.

Microsoft founder Bill Gates could have devoted his vast, self-made fortune to anything that interested him. What he chose to do, with his wife, Melinda, was create a foundation based on a simple premise: "All lives have equal value." Through healthcare, poverty alleviation, and education, the Bill & Melinda Gates Foundation has taken a leadership role in putting that premise into action, and for that, we believe, the Gateses deserve the Nobel Peace Prize. We can't think of anything that could contribute more to world peace than giving each child an equal opportunity to grow up healthy and with the tools necessary to make his or her way in the world.

Despite their significant interest in pandemic preparedness and outbreaks that could cause millions of deaths in a short period of time, the Gateses are focusing on the basics in a way that could make a huge difference, worldwide. "That's what the foundation spends most of its time on when it comes to health," Bill Gates noted. "We're not an epidemic, bioterrorism defense organization. We're a malaria, HIV, TB, diarrheal disease, pneumonia organization."

One of the foundation's major efforts is a heroic assault on polio. I admit to a long history of skepticism about the possibility of completely ridding the world of polio, particularly in light of today's fractured and vulnerable nations and the political, economic, and religious issues attached to them. But it looks like it might finally happen, thanks to actors like the Gates Foundation and the partners it has inspired.

Even more important, however, is the assault Gates is waging on malaria and how he is engaging partners worldwide to get involved.

Though we think of polio as a more "emotional" disease than malaria, primarily because we have suffered through it in the Western world and recall the pitiful images of children in leg braces, wheelchairs, and iron lungs, it may actually be the "easier" disease to conquer. Just like smallpox, it is species-specific to humans, with no animal reservoir and no mosquito vector. Malaria is a different story.

Malaria has been around throughout recorded history, and the two most effective drugs—quinine and artemisinin—are derived from ancient remedies: cinchona tree bark and the *qinghaosu* plant. It is caused by a parasitic single-cell microorganism (a protozoan) called a plasmodium that is transmitted by the *Anopheles* mosquito. As we will detail in chapter 14, this mosquito is very different from *Aedes,* the vector for dengue fever, yellow fever, Zika, and chikungunya. Control efforts for each species of mosquito must come from separate playbooks, based on where they live, breed, and feed.

Once the malaria plasmodia enter the bloodstream with the mosquito's saliva, they travel to the liver and reproduce. Symptoms of malaria include high fever, nausea, vomiting and diarrhea, sweating or shaking chills or both, fatigue, and headaches. Because of liver involvement, jaundice may develop. Severe cases can result in encephalitis, breathing problems, and anemia, which, in turn, can lead to coma or death. Those already at a disadvantage due to endemic poverty, unclean water, inadequate medical facilities and support, and other health issues are more likely to suffer severe forms of the disease. Once a person is infected, the infection can be transmitted person-to-person through blood transfusions, through sharing a needle, or from mother to unborn child. Unlike many of the infectious diseases we have discussed, malaria can recur. In children, it can cause lifelong intellectual and learning difficulties.

So important is the fight against malaria that it has resulted in five Nobel Prizes in Physiology or Medicine, from 1902 to 2015. On the other hand, plans to eliminate the disease worldwide were abandoned in 1969 as being too expensive, complicated, and impractical.

Malaria is present in about a hundred countries, with about 90 percent of the deaths occurring in sub-Saharan Africa. Seventy-seven percent of those deaths are among children under five.

With the involvement of the Gates Foundation and others, cases dropped about 25 percent from 2004 to 2016, with deaths falling by 42 percent. During that time, malaria funding increased nearly tenfold and major gains have been made in controlling the disease in developing nations. This success has been the result of a combination of interventions, including timely diagnosis and treatment, indoor spraying with effective agents, and bed nets impregnated with long-lasting insecticides. The Gates-supported Global Fund to Fight HIV, Tuberculosis and Malaria is the largest purchaser of bed nets.

In 2013, the Gates Foundation announced a new multiyear malaria strategy called "Accelerate to Zero." I was initially skeptical of the foundation's conclusion that the eradication of malaria is biologically and technically feasible. But after Mark and I talked to Gates about this initiative we came away with a sense of admiration for his "We won't know until we try" mind-set. As he said to us, "These things don't come in black-and-white form. And the time when you have to act is when it's not that clear."

We have learned the hard way that once resources become scarce for vector control—which invariably happens over time—the mosquito populations and the viruses they carry quickly rebound. And even if we can eradicate them from one continent, we must be forever vigilant that they don't reenter via plane or ship from another infested area. Global eradication has to be the ultimate goal. Frankly, if anyone can pull off this effort in my lifetime, it will be Bill and Melinda Gates. It would be an amazing legacy gift to humankind.

The strategy involves several fronts, all of which entail making sure malaria has a prominent place on the global health agenda. Two of the most important elements fall in the preventive phase: the development of new insecticides against the *Anopheles* mosquito and work on a vaccine. At present, more than thirty malaria vaccines are in some phase of development. After five years of development by the National Institute of Allergy and Infectious Diseases, one candidate vaccine has produced encouraging results in its first trial in humans.

Genetic means of releasing sterilized mosquitoes into the wild are being tried with several dangerous vector species. The ultimate effectiveness of this technique is still highly speculative, and scientists are still trying to figure out how to give the modified male mosquitoes a selective advantage over their "natural" counterparts. Also, since nothing on this order has ever been attempted, there are concerns about unforeseen and unintended consequences to the ecosystems in which they are introduced. Some experts predict it will take ten years to know if this strategy will work.

With vector-borne diseases, we make a distinction between active and passive measures. Active measures include insecticides to kill the insect carriers and pharmaceuticals to treat the diseases and symptoms. Passive measures include bed nets. One of the more interesting passive measures being tested is insecticide-treated wallpaper. Insecticide spraying must be repeated every three or four months, but these wall liners can be effective for three years or more.

The US Army has been issuing combat uniforms embedded with the synthetic insecticide permethrin to personnel in mosquito-endemic regions for several years. Experiments are now under way to see if clothing treated with insecticide might be effective protection for civilian populations in afflicted areas.

On the treatment side, the Gates Foundation is supporting what it refers to as a "single-dose cure: a pill that would wipe out all parasites in the body." Existing drugs, to which the disease is developing resistance, have to be taken for three days, so many people do not finish their doses.

These efforts dovetail with the President's Malaria Initiative (PMI), which was launched in 2005 after the 2003 passage of the Global Leadership Against HIV/AIDS, Tuberculosis, and Malaria Act, which was amended in 2008. With a goal of reducing malaria-related mortality by 50 percent, the initiative has aimed to scale up four specific efforts: providing and making insecticide-treated nets more effective, indoor spraying, artemisinin-based combination therapies, and intermittent treatment of pregnant women.

Sustainability, as you will have gathered by now, is always a prime issue in public health. But what will happen when, as we hope and expect, the effort and resources being expended on African malaria will cause case numbers to continue to fall? Will the fight no longer seem urgent? Will we move on to the next pressing thing, as we did with Ebola and mosquito control in general? Or will we see the effort through, as we did with smallpox, to the lasting betterment of the world?

HIV/AIDS was one of the most reported and tragic stories of the 1980s and early 1990s. The gaunt faces of those waiting to die from this incurable infection continue to haunt the memories of anyone who lived through those times. With the remarkable progress in antiretroviral therapy—though still no effective preventative vaccine—the affliction has transformed from a near certain death sentence to a manageable chronic disease—at least in countries wealthy enough to afford the treatment or fortunate enough to be the recipient of international aid.

But with progress has come a retreat from the headlines and a measure of complacency about a disease that is still a major world problem.

This is the state of HIV/AIDS in the world today:

There are about 2 million new infections each year, and sub-Saharan Africa accounts for almost 70 percent of them. Around 220,000 of these new cases are in children younger than fifteen, most of whom were infected by HIV-positive mothers in utero or through breastfeeding. About half of those living with HIV don't know they have it. Most people living with HIV, or at risk for HIV, don't have access to prevention measures, care, or treatment.

A few African countries, most notably Kenya and South Africa, have made significant strides in getting treatment to portions of their afflicted populations. But for much of Africa and the Middle East, nothing is being done for the majority. Some people who know they are HIV-positive are told by health workers to come back for treatment only after they develop symptoms, because resources will only stretch to those with active disease. Many are unwilling to come forward because of job discrimination, social ostracism, or religious persecution in countries like Nigeria, Uganda, and Russia. In some places the distribution of condoms and clean needle exchanges can help stem the tide. In other places, they also are targets of local social taboos.

The United Nations has set a target date of ending the AIDS epidemic by 2030. But at the UN High-Level Meeting on Ending AIDS in June 2016, the delegates could agree on everything except how to achieve that end. Their declaration supported the WHO guidelines that every HIV-positive individual have access to treatment and recognized the consequences of coming up short of the goal.

But some members wouldn't accept the inclusion of language in the document about gender equality and access to HIV prevention and contraception for women. "This runs counter to the legal framework of several countries," commented the representative of Sudan. Some didn't like language that urged sexual education to prevent transmission. Others thought it distasteful to single out vulnerable groups such as IV drug users, sex workers (a term Iceland didn't care for), homosexual and transgender people, and prisoners. Iran's representative went so far as to say this language was discriminatory. The Vatican, a nonvoting member, took exception to any mention of birth control measures, and others wanted more emphasis placed on abstinence before marriage and fidelity during marriage.

The US representative, Sarah Mendelson, said her nation believed the document "should have been stronger and more explicit" on human rights, reproductive rights, and marginalized populations. The representatives from Canada and Australia agreed, decrying the absence of language calling for the end of anti-gay discrimination and stigmatization.

All this disagreement doesn't bode well for an action plan to defeat the disease.

The country that has done the most to combat AIDS worldwide is the United States, through PEPFAR—the President's Emergency Plan for AIDS Relief, spearheaded by President George W. Bush as a means to provide treatment for millions in resource-limited areas and prevent spread. It was renewed and expanded in 2008 as the cornerstone and largest component of President Barack Obama's Global Health Initiative and represents the greatest and most extensive health initiative a single nation has ever launched against a single disease. Multiple government agencies are involved and coordinated, including the Departments of State, Defense, Health and Human Services, Commerce, and Labor, as well as the CDC, the US Agency for International Development, and the Peace Corps. PEPFAR now works directly with host countries to develop local leadership and long-term sustainability.

The overall goal is that the programs become country owned and country driven as part of an overall plan to address local health needs. As PEPFAR transitions from an emergency response to an ongoing effort, the aim is to develop the local expertise necessary for evidence-based decision making. And like the Gates Foundation, PEPFAR also wants to leverage the work of multinational organizations and international partnerships.

As a US citizen, I am very proud to see what PEPFAR has accomplished in reducing the global burden of HIV. But I'm worried about the future impact that this program will have.

First, given the current level of political support—or lack thereof—for public health–related issues like our response to the Zika outbreak, there is no guarantee that funding for PEPFAR will remain at its current level. In fact, since that major increase in PEPFAR support in 2008, federal funding has flatlined and will be lower for fiscal year 2017 than it was for fiscal year 2016.

This despite an ever-growing number of persons living with HIV infection. In 2010, there were an estimated 33.3 million HIV cases globally. By 2015, that number had increased to 36.7 million, a net gain of more than 3.4 million. In 2015, PEPFAR provided antiretroviral therapy to 9.5 million HIV cases. If the rate of increase in new persons living with HIV infection continues and they all need ongoing treatment, within a decade there will be 6.8 million new cases. That's 71 percent of the number that are currently being treated via PEPFAR support today, meaning we would need a substantial increase in the support for PEPFAR over the next decade just to keep pace with the new cases. I don't see that happening unless the governments of the world where new HIV cases are occurring step up and provide this support. Given that almost half of the global increase in new HIV cases is occurring in Western and Central Africa, the odds are distressingly slim.

The best answer to this situation is to find an effective vaccine or curative treatment like we have for hepatitis C virus infection. But that hasn't happened and isn't happening, though not for lack of trying. Nearly a billion dollars is spent on AIDS vaccine research every year. Dr. Tony Fauci, director of the National Institute of Allergy and Infectious Diseases, has been involved with HIV and AIDS since the beginning. "It's a scientific dilemma," Tony explains, "because the body doesn't like to make neutralizing antibodies against HIV, and we're going through all kinds of the most eloquent science you can imagine: using cryoelectron microscopy, structural biology, and X-ray crystallization to get the right conformational form of the envelope to engage germ line B cells to induce a protective response; I mean, all kinds of highly sophisticated efforts."

I still don't know if we'll see an effective vaccine anytime in the near future, though I am hopeful. At the same time, I think we have to plan our ongoing war against HIV/AIDS without such a "nuclear" weapon. We have to think of that war as a series of ongoing local battles.

Tuberculosis doesn't engender in us the same level of alarm as newly emergent infections, but it should. Though we think of it as a vestige of the nineteenth and early twentieth centuries, evoking mountaintop sanitaria and opera heroines coughing up blood, tuberculosis is very much with us in the world today, and more and more of it is becoming drug resistant. For a long time it had become rare in developed nations, but it came back around the same time as HIV, and in many areas of India and the developing world, there is substantial comorbidity with HIV, complicating treatment options enormously.

Tuberculosis is caused by a bacterium that can affect various parts of the body but most commonly infects the lungs. It is spread person-to-person through the air, though fortunately it is more difficult to pick up than many respiratory viruses, such as measles and influenza.

In otherwise healthy people, TB may not cause any symptoms because the immune system can wall it off; that is, live TB bacteria reside in the body but are in nodules walled off by immune cells. The WHO estimates that as much as one-third of the world's population may be infected with latent TB. People in this category have about a 10 percent lifetime risk of the disease becoming active. "Active" TB causes symptoms such as coughing (sometimes with blood), chest pain, weakness, weight loss, fever, and night sweats.

But if someone who has latent TB is also infected with

HIV-related disease, all bets are off. The combination of TB and HIV becomes an infectious-disease perfect storm. The immune system of the person with HIV is compromised, giving the TB bacteria free rein to grow and spread through the lungs or whichever organ the bacteria is residing in. These patients often have a lot of damage in their lungs from the TB bacteria and thus are much more highly infectious to others. One of my most challenging investigations as Minnesota state epidemiologist was following up on hundreds of air passengers who had taken a flight to Minneapolis–Saint Paul from some distant country, only to find out later that a passenger on their flight had active drug-resistant TB and was infected with HIV. He coughed for nine hours on the flight to the Twin Cities.

Dr. Aaron Motsoaledi, the charismatic and highly respected chief health minister of South Africa, has been outspoken in trying to warn the world of the renewed threat posed by tuberculosis, which, without treatment, kills about 45 percent of its victims. He points out that 4,100 people die every day from the disease. And yet this is one of those examples of our emotional disconnect regarding the most likely threats. We are terrified of Ebola but ignore TB on the same continent. And make no mistake; TB is a much more likely large-scale killer in the West than Ebola or Zika.

Motsoaledi called together leaders of the mine workers and other important unions and pointed out the facts: In 2009, eighty people died in South African mining accidents and there was outrage. That same year, 1,500 miners died of TB and it was as if no one even noticed.

The TB death, he told the *Huffington Post,* is "a process, not really an event. It happens very slowly, maybe in a corner somewhere, in an isolated hospital ward, with nobody watching, so it doesn't evoke any emotion."

The good news is that we have made some real inroads in impacting global TB mortality over the past fifteen years, reducing it by 47 percent. The bad news is that in 2014, only 6 million new cases of TB were reported to the WHO, fewer than two-thirds (63 percent) of the 9.6 million people estimated to have become ill with the disease. This means that worldwide, almost 40 percent of new cases went undiagnosed or were not reported. It's unclear if these infected individuals had access to appropriate healthcare. And there is more bad news: Of the 480,000 cases of multidrug-resistant TB estimated to have occurred in 2014, only about a quarter of these—120,000—were detected and reported.

These numbers are why it is important that organizations and government agencies such as the Gates Foundation are committed to research in the TB area. Specifically, Gates has funded work in three areas: vaccine development, rapid diagnostics, and new drugs to combat resistance. But for the Gates investment to pay real dividends, it will have to be seen as a leadership example that other organizations and governments can model themselves on and take an active role in.

With proper care and treatment, tuberculosis remains a curable disease in most cases. But we've mentioned the increasing multidrug-resistant strains in chapters 16 and 17, on antibiotic resistance. With the most highly antibiotic-resistant strains of TB, successful treatment is not a guarantee, even with modern high-tech medicine. Unless we can get out in front of it, the disease will always be another river flowing against us at five miles an hour faster than we can swim.

The combination of major population growth among those living in the crowded, squalid conditions of the megacities of the developing world, the extensive global movement of people around the world, and increasing TB drug resistance makes for a very dangerous TB future for all of us. Support for TB prevention and control needs additional investment, not less. If we don't pledge ourselves to this, I'm certain we will more than pay for it in the long run.

CHAPTER 10

Gain of Function and Dual Use: The Frankenstein Scenario

You seek for knowledge and wisdom as I once did; and I ardently hope that the gratification of your wishes may not be a serpent to sting you, as mine has been.

— Mary Shelley, *Frankenstein*

At the end of Mary Shelley's famous novel, scientist Victor Frankenstein explains to his new confidant, Arctic explorer Robert Walton, that scientific adventurism is a double-edged sword, and the same labors and discoveries can have opposite effects, depending on how they are handled, and by whom. Frankenstein tells Walton that though his own scientific advances have created nothing but misery and destruction, perhaps others who follow may create healing and progress.

A careful reading of *Frankenstein* reveals that the reanimated and quickened body of dead flesh did not become a monster through its own inherent evil, but because of the way its creator, and others, reacted to it.

This is the cautionary tale with which we approach the subject of gain-of-function research of concern (GOFRC) and dual-use research of concern (DURC).

Gain-of-function, as you will remember from chapter 4, is

an intentionally created mutation through one of several methods, which gives the microbe new functions or abilities. DURC is life-science research that could be directly misapplied and pose a significant threat to public health and safety.

By now, it should be evident that one of the underlying themes in our understanding and response to infectious diseases in the twenty-first century is the power of microbial evolution. As we described in chapter 5, evolution is the force that drives diversity, and it is based on the concept that the most fit survive. The modern world guarantees that evolution will change the microbes we live with, particularly as they have an opportunity to infect the billions of people on the earth now, compared to the millions that existed just a century ago. The same holds true for the increasing number of animals, particularly those related to agricultural production. The fact that microbes, like their animal and human hosts, jet around the world with frequency and speeds that never occurred before means that they will spread quickly to the most distant parts of the globe. All of these factors favor the emergence of microbes that will survive and even thrive in spite of the control measures, vaccines, or treatments we employ in our battle against them.

Now we have the potential for creating hyperevolution: the kind of change in microbes that Mendelian genetics or Darwinian evolution could not necessarily have predicted.

This type of evolution occurs as the result of microbial engineering, a human activity that intentionally manipulates the genes of a microbe to fast-forward evolution by thousands of years or, in some instances, create changes that evolution might never have achieved. One example is a generic chimera agent, named for the mythical fire-breathing creature with the head of a lion, the body of a goat, and the tail of a snake. Some new live virus vaccines are just that—taking parts of one virus and inserting it in another live, replicating virus. This is possible only because of human intervention: swapping and exchanging genetic material from multiple microbes. Chimeras can be created for both useful and nefarious purposes.

How does this new evolution model affect the risk of infectious diseases in the twenty-first century? It's all about the rapidly growing power of technology.

In 2007, Steve Jobs introduced the first iPhone to the world. That was just ten short years ago. The power of the current iPhone dwarfs that of the original model. During those same ten years, life sciences, and specifically microbial genetics, have undergone a similar revolution of capability and power. We now have microbiology tools to manipulate the genes of microbes that twenty years ago may have been available to only our most advanced government labs. Yet today they are in our high school microbiology classes and used by amateur DIY scientists. Could a genetically manipulated microbe be transmitted to a human or animal and cause disease? It's a real possibility. We need look no further than the recent debate about both the exciting promise, and disturbing peril, of gene-drive technology.

One exciting new genetic engineering tool involving potential GOFRC is CRISPR, which stands for clustered regularly interspaced short palindromic repeats and refers to a DNA sequence that repeats at regular intervals in about 40 percent of bacteria. Researchers are now using CRISPR to "edit" DNA to produce more desirable versions of various plant and animal species. It may be possible in the not-too-distant future to use CRISPR to create whole new species.

Compared to older gene-editing techniques, CRISPR is much cheaper, simpler, and faster, with the potential to create a whole new range of genetic modifications. The promise that this type of research holds for attacking the most serious infectious diseases of our time is exciting. At the same time, it is not difficult to imagine what could happen if this increasingly available technology were used for diabolical purposes. In his February 2016 "Worldwide Threat Assessment" testimony to the Senate

Armed Services Committee, James R. Clapper, the director of national intelligence, said that gene editing had become a global danger.

DURC is not a new issue. In the early days of atomic physics, the scientific research community realized this work could be used to bring both benefits and harm to society. After World War II, the threat of biowarfare, the use of infectious agents to intentionally harm the military and general populations of an enemy, did not involve your standard microbiology researcher in academia or in organizations like the NIH or the CDC. Rather, the work was framed as research with both civilian and military applications. Often it was classified and was conducted in military research laboratories where public dissemination of the methods or results of the studies was never intended.

Only after the events of 9/11 and the subsequent anthrax attacks in the United States did both the government and the scientific community take seriously the potential for harm that could result from DURC. Meanwhile, the exploding revolution in life sciences continued.

In 2004, Professor Gerald Fink of the Massachusetts Institute of Technology chaired a now historic National Research Council Committee that produced a document commonly known as the Fink Report. It set the stage for how to consider ways to minimize threats from biological warfare and bioterrorism without hindering the progress of biotechnology. It is generally agreed throughout the life sciences that biotechnology is an essential part of any modern solution for improving global health. The Fink Report summarized the response of the life sciences community to the increased concerns about bioterrorism. It also concluded that DURC should not be prohibited but should be scrutinized carefully and undertaken only with the awareness of potential misuse.

The final Fink Committee Report made seven overarching recommendations, including the need for HHS to augment the already established system for review of experiments involving recombinant DNA and to create a review system for seven classes of experiments, labeled "Experiments of Concern."

The report also called for the creation of a national scientific board to provide advice, guidance, and leadership for the system review. That board was established in 2004 as the National Science Advisory Board for Biosecurity (NSABB). The NSABB was made up of twenty-five voting members representing key stakeholder perspectives to provide expertise in a variety of areas related to microbiology, infectious diseases, laboratory biosafety and biosecurity, public health, and bioethics, and eighteen ex officio members from various federal agencies.

In the summer of 2005, I was appointed to the NSABB as a charter member by HHS secretary Michael Leavitt. I don't think any of us on the board had a clear idea of what our immediate agenda should include. That changed when we were suddenly handed a hot-potato issue. The CDC and three other research groups submitted a paper for publication in the journal *Science* detailing how they had reconstructed the 1918 H1N1 influenza virus, using virus genes that had been identified in lung samples of patients who died during the 1918 pandemic. From that information, they were able to re-create the virus and then put it into ferrets (a good animal model for human influenza infection) to understand how easily it could transmit, how it causes illness, and its severity. The researchers' primary questions were: How did the pandemic virus evolve and adapt to humans? Could the new reconstructed virus identify mutations to be used for surveillance? Why was the virus so deadly, especially in young adults? Could this data be used to develop novel drugs and vaccines?

Secretary Leavitt sent the paper to the NSABB, asking the board to determine whether it should be published in the general medical literature. The central question was, If others were able to re-create the work, would that influenza virus pose a serious public health risk if it were to be released accidentally back into the general population?

We were rather ill prepared to address this question; we had no standardized criteria or protocols for determining the risk that this virus posed to the general public's health. At the time, it was believed that the virus posed little additional risk to the general population, a large portion of which had been exposed to the H1N1 influenza virus during the twenty-five years it had circulated as a seasonal flu virus. After several conference calls and a full board meeting, we agreed that the paper could be published with the provision of some additional information on how to reduce the risk of the virus being accidentally released from the laboratories where the work was done. In retrospect, we now know that infection from a previous strain of H1N1 offered no immunity or protection against the 2009 H1N1 influenza pandemic strain that emerged from Mexico four years later. In fact, studies showed that most people would have been just as susceptible to the reconstructed 1918 pandemic virus.

This experience provided two valuable lessons. First, our assumption that infection from any recently circulating H1N1 strain would have provided protection from the devastating 1918 H1N1 strain was wrong. Second, it was a wake-up call that these artificially constructed viruses posed the potential for globally catastrophic effect. This was no longer theoretical. We were confronted by scientific reality.

A few years later we faced a similar challenge, with higher stakes. In the fall of 2011, two manuscripts were submitted to science journals summarizing research on the virulence of mutated influenza virus H5N1. The research was supported by the NIH and conducted by Professor Ron Fouchier and colleagues from Erasmus Medical Center in the Netherlands, and by Professor Yoshihiro Kawaoka and colleagues from the University of Wisconsin–Madison.

H5N1, considered the grandfather of bird-flu viruses, has

been a serious public health concern since it was first identified in 1997 in Asia and has had a devastating impact on domestic and wild bird populations there. Humans can be infected with it following exposure to infected birds. While it rarely infects humans, when it does it causes severe disease with case fatality rates of 30 to 70 percent. However, to date it has not successfully maintained the ability to be transmitted by infected humans to other humans.

We were confronted in this case by a powerful real-world example of DURC. The two studies successfully created forms of H5N1 that could be transferred between ferrets via the respiratory route—that is, through the air. The purpose of the research was to determine the possibility of predicting which genetic changes corresponded to an avian influenza virus, such as H5N1, becoming readily transmissible between mammals. We can't say for sure that what happened with ferrets would happen with humans, and we certainly didn't want to find out. But this suggests a plausible and very scary possibility.

The NSABB was asked by the US government to assess the DURC implications these manuscripts created, and I was asked to serve on a working group to review this work and make recommendations to the full board as to the potential harm of publishing these data. As with the H1N1 research five years earlier, the question was, Would publication of the methods and findings of this work allow others to create potentially transmissible influenza viruses in humans that also had an increased ability to cause serious, life-threatening disease?

At that time (and as of today), transmission of H5N1 virus from humans to humans has been rare. However, it remains one of the bird-flu viruses that have the potential to become a human pandemic strain. If H5N1 influenza virus acquired the capacity for human-to-human transmission and an increased case fatality rate, we could face a worldwide pandemic of devastating impact.

The NSABB debated the benefits of publishing such research, including the possibility that if similar viruses were found circulating in bird populations, we would have early warning of a potential pandemic. After several months of conference calls and document sharing, the working group concluded that these scientific findings represented a grave concern for global biosecurity and that their communication should be limited, meaning only a very general, high-level research manuscript summarizing the methods and results should be published. This recommendation, a highly unusual one for such work in the life sciences, was then considered by the entire NSABB and reaffirmed with a unanimous vote. I believe it represented a careful consideration of both the potential benefits of publication and the potential harm that could occur from such a precedent. Along with our recommendation to limit the communication of the results, we encouraged a rapid and broader international discussion of DURC research concerning H5N1 influenza viruses with a goal of developing a consensus on the path forward.

This was not the end of the matter. Researchers from both sides of the publication issue continued to debate the appropriateness of the NSABB recommendation to the government. Those supporting full publication of the research reiterated that other experts, funders, and external reviewers had been supportive of the need to identify viral factors that affect transmission and contribute to the emergence of pandemic viruses. These studies supported those efforts. They wrote that the risk to the public and the environment had been reduced to "the absolute minimum," arguing that strict biosafety measures were in place to protect the researchers, the environment, and the public, adding that even in the remote chance of human error in the lab setting, workers had access to H5 vaccines and antivirals and could be quarantined if exposed. That group, in favor of uncensored publication, also argued against withholding full details of the studies, claiming that techniques to create airborne viruses are well-known and the transfer of the virus to high-containment laboratories was not necessary. They concluded that "censoring the manuscripts on A/H5N1 virus transmission will, therefore, only create a false sense of security."

On the opposite side, I joined several colleagues in publicly explaining why we had grave concerns about the studies' being published. We argued that transmission of the influenza virus between ferrets didn't mean the mutated virus would spread among humans or other mammals, but that possibility could not be excluded. Thus, the publication of the full study details could make it easier to reverse engineer the virus and actually make a mutant strain that could be transmitted.

We were worried that intentional or accidental release, even if the virulence of the virus was similar to that of the wild-type H5N1 viruses, could boost the number of human cases and pose a threat that the virus could swap genes with other influenza viruses and create a new pandemic strain. Finally, we urged that decisions about research that carry significant risk to the public should not be made by life scientists alone and should include input from scientists without conflicts of interest, including biosecurity experts from outside the life sciences community.

There was pressure mounting to overturn the NSABB decision from various life sciences research groups concerned about the precedent the censoring of this research created. The NIH, which provided the financial support for the studies in question, urged the NSABB to take another look at the issue. NIH director Dr. Francis Collins maintained that because of specific provisions in US government export-control requirements, either all or none of each manuscript could be published. The US government controls the export of sensitive equipment, software, and technology as a means to promote US national security interests and foreign policy objectives, and this H5N1 research met export-control requirements for regulation. Most of the NSABB wanted a redacted manuscript to be published for the purpose of alerting the world of this new development. Now the board was being told to publish either everything or nothing.

The NSABB was reconvened on March 29–30, 2012, at which time the US government requested that we reconsider our previous decision recommending the redaction of both manuscripts before publication. It was clear to me and a number of my colleagues that the leadership of the NIH wanted us to approve full publication of both articles. We didn't believe there was a sinister motive for their request that we find a solution for full publication of these papers. But I do believe there was a bias toward finding a solution that was less about risk-benefit analysis and more about how to get the NSABB out of a difficult public policy situation.

The board was also provided with information suggesting that this work could help quickly identify an emerging pandemic strain of virus and allow for earlier efforts to secure a pandemic vaccine. However, through my extensive work on influenza, I knew that this was not true. In the end, the board revoted and approved the full publication of both manuscripts. I walked out of the boardroom that day feeling as if I had just played a crazy game of public policy Jeopardy: Here is the answer, now help me identify the right question that supports it.

The sobering lesson for me from the H5N1 manuscript debate is that weighing potential benefits against clear risks for pathogens of pandemic potential is extremely complex and difficult to control. Like climate change and antimicrobial resistance, this is an issue that goes far beyond our borders. For one thing, substantial DURC and GOFRC research is conducted throughout the world by individuals who, for reasons of mental instability or criminal intent, wish to harm large numbers of people. Then there are the irresponsible academic, corporate, or amateur scientists who simply don't fathom the risk their work potentially poses.

So the issue of whether this work should be done at all comes down to two key questions: Does the work have legitimate scientific purpose? Can the work be done safely in a lab, protecting both its own workers and the members of the community? Then, if the work is worthwhile and can be conducted safely, should it be fully exposed to the public, including methods and results, in the form of publication in a medical journal?

The following story is not about GOFRC organisms specifically but is a real-world example of what can happen when a bug, by whatever means, escapes the confines of the lab where it is being studied or developed.

Prior to 1977, it was accepted that the emergence of a new pandemic strain would result in the disappearance of the previous seasonal flu strain. After the infamous 1918 influenza pandemic, the new H1N1 virus became the seasonal virus for the years following. The seasonal H1N1 virus was much attenuated compared to its pandemic ancestor, and many had immunity against it after becoming infected during the 1918–19 pandemic. Then, in 1957, H2N2 emerged as the next pandemic influenza strain. In the following months, H1N1 disappeared and H2N2 became the new seasonal circulating flu virus. It happened one more time, in 1968, when the H3N2 influenza virus caused the next pandemic and, shortly thereafter, H2N2 disappeared. Based on how H1N1 had evolved after the 1918 influenza pandemic, we thought we could count on only one strain of influenza A virus circulating each season, even though we couldn't explain why.

That all changed in 1977, when an H1N1 influenza virus appeared in Asia and quickly spread around the world. It did not displace H3N2. Now we had two cocirculating seasonal influenza strains.

How had this happened? When our team was researching information for the 2012 CIDRAP report, *The Compelling Need for Game-Changing Influenza Vaccines,* we uncovered documents long forgotten in the federal files that addressed the appearance of

H1N1 in 1977. The virus showed up almost simultaneously in eastern Russia and western China in May of that year. The genetics of the virus showed that it was very similar to the H1N1 that disappeared in 1957, twenty years before. Had the virus been circulating in nature for all those years, the genetic makeup of the virus would have been very different. It was apparent that the new virus had sat in someone's freezer for twenty years before making its return to humans.

It turned out that the Soviets were conducting vaccine studies using live, attenuated H1N1 influenza viruses in the very area where the new H1N1 was first detected. During our research, we uncovered a letter from the Soviets to the US government requesting that we share with them the 1976 Fort Dix strain of H1N1 for their vaccine studies. I have little doubt that the appearance of the 1977 H1N1 virus and its rapid global transmission in just several months was the result of a release of the virus in the course of the Soviet vaccine studies.

We don't know exactly what they were doing with the virus. What we do know is that it got out, either accidentally or on purpose, causing a local outbreak in lab workers that subsequently spread around the world. Either way, the powerful lesson here is that if an influenza virus accidentally escapes or is intentionally released, expect that it will spread around the world in short order. This is the proverbial single match being able to light a global forest fire. The possibility for a DURC research study using a potentially dangerous influenza virus should scare the hell out of everyone.

Over the past five years, both the CDC and academic labs around the world have documented accidents where a variety of pathogens were, or may have been, released. Fortunately, most of these did not put the general public at risk, but they could have. If this can happen at a place like the CDC, where some of the leading lab experts in the world work in state-of-the-art facilities and where the spotlight makes it likely that the general public will find out about such problems, imagine what could happen in the thousands of other labs around the world. If we are going to do DURC research studies that involve microbes like influenza, there is zero margin for error.

Does that mean we shouldn't conduct such research? During the H5N1 debate, I was struck by the rigidly black-and-white positions that so many of my colleagues took. There were those who strongly believed it should be done as the researchers proposed—kind of an academic freedom issue—and there were those who believed it should never be done, as if it crossed some kind of moral line.

I felt out of step with this black-and-white logic then, and I still do today. I believe that work like the H5N1 studies can provide us with unexpected, game-changing results. Knowing whether Ebola virus can become a respiratory-transmitted pathogen would certainly be a game changer, for example. There are other DURC research projects I want to see done that also would be potentially dangerous if there was an accidental release or the methods and findings of the studies were made fully available in the scientific literature, thus enabling others who have nefarious intent, or who employ lab safety practices that make a release a high-risk possibility, to do the work.

The answer is clear. We need to do these studies in a few select labs with leading experts and state-of-the-art safety features. And this research needs to be classified—or at least considered sensitive—so that the results are shared only with those with a need to know. We can support the US government and other responsible governments of the world in anticipating and preparing for potential microbe-related crises with this approach.

In 2016 the NSABB completed a two-year process to finalize comprehensive recommendations for the US government in assessing and funding GOFRC studies on H5N1 and other pathogens of pandemic potential. It reflects a major improvement over the information we had to work with in 2012. However,

I believe that there are still serious issues in the new NSABB document, *Recommendations for the Evaluation and Oversight of Proposed Gain-of-Function Research,* which describes the NSABB's seven major findings and seven linked recommendations.

The finding I have most trouble with relates to when to do, or not do, GOFRC work. The NSABB concluded, "There are life sciences research studies, including possibly some GOF research of concern, that should not be conducted because the potential risks associated with the study are not justified by the potential benefits."

If we were to adopt the classified research model, with the highest possible levels of lab safety employed, we should be able to engage in any GOFRC research if there is possible benefit to our preparedness for early recognition or to our response to either a natural or man-made microbial catastrophe event.

But let's not fool ourselves. While the life sciences community and governments can be the first wall of protection against either nefarious deployment or lab safety issues associated with DURC or GOFRC, we have to be realistic that we're not going to catch every one. As the Irish Republican Army is reputed to have said, "You have to be lucky all the time, we only have to be lucky once."

I commend the NSABB for identifying the need to engage the rest of the world, including nongovernmental organizations and private companies. If another event such as an H1N1 release occurs in a foreign country, we'll still be deeply in the soup. So we must bring all governments to the table to enlist their support and action in this area.

Of all the concerns addressed in this book, DURC and GOFRC may be the most troubling in that we have, as of now, no satisfactory answers or solutions. The technology to conduct this work will only get more sophisticated and accessible in the years ahead. In this Internet age, it is probably unreasonable to expect a complete and impenetrable security blanket over critical scientific findings. We just have to do the best we can.

CHAPTER 11

Bioterror: Opening Pandora's Box

Half fearfully and half eagerly she lifted the lid. It was only a moment and the lid was up only an inch, but in that moment a swarm of horrible things flew out. They were noisome, abominably colored, and evil-looking, for they were the spirits of all that was evil, sad and hurtful. They were War and Famine, Crime and Pestilence, Spite and Cruelty, Sickness and Malice, Envy, Woe, Wickedness and all the other disasters let loose in the world.

— "The Myth of Pandora," interpreted
by Louis Untermeyer

On October 4, 2001, I was in the CBS studios of *60 Minutes* in New York to talk about my book, *Living Terrors: What America Needs to Know to Survive the Coming Bioterrorist Catastrophe.* It had been published more than a year earlier to middling sales, but following the horrific 9/11 attacks my subject had abruptly become disturbingly relevant. Mike Wallace was the reporter-interviewer on the story. The three other guests with me were Dr. David Franz, another one of my mentors on bioweapons and a colonel who had headed up USAMRIID (the US Army Medical Research Institute of Infectious Diseases), former ambassador Richard Butler, lead UN weapons inspector, and Dr. Matthew Meselson, the Harvard molecular biologist who had studied under double Nobel laureate Dr. Linus Pauling.

Suddenly, executive producer Don Hewitt rushed into the studio and broke into the interview, grasping a news bulletin. "Tell me what the hell you know about this anthrax case!" he demanded of the four of us. Just moments before, Florida health officials had announced that Robert Stevens, a photo editor who worked for the *Sun,* a supermarket tabloid, had been diagnosed with pulmonary anthrax, the first case in the United States in almost twenty-five years. Mr. Stevens died the next day.

As it happened, we knew nothing about the case, though we would all become deeply involved in the coming days. The trillion-dollar question: Was this an isolated case with some environmental exposure to an infected animal, or was this the first shot of an attack? Anthrax had always been a prime candidate for use as a bioweapon. Coming so close on the heels of the 9/11 attacks, it seemed highly unlikely that if additional cases of anthrax were found, this outbreak could be accidental.

A week later, I was in Washington meeting with HHS secretary Tommy Thompson's staff discussing the unfolding anthrax crisis. By that point, letters containing the highly lethal anthrax powder had been received by four other news media organizations on the East Coast: ABC, CBS, and NBC News and the *New York Post,* in addition to the *Sun*'s publisher, American Media, which also published the *National Enquirer,* among other similar periodicals.

Minnesota senator Paul Wellstone, who would tragically die in an airplane crash almost exactly a year later, was aware I was coming to Washington and asked if I could also brief his staff and the Senate majority leader, Tom Daschle, on the situation while I was there.

I will never forget meeting with the senators in Daschle's ornate Capitol office, explaining to them how the letters containing anthrax powder caused the human disease they were unleashing. I also noted that based on the quality of the powder, whoever was perpetrating this horrible event likely had additional stock that had not yet been mailed. Five days later, Senator Daschle's office in the Hart Senate Office Building received the first letter containing anthrax powder sent to an individual outside the media. The same day, a letter addressed to Senator Patrick Leahy of Vermont arrived in Washington. The letters were crudely written condemnations of the United States and Israel and proclamations of Allah's greatness. This had now become an out-and-out attack against our federal governmental institutions.

Altogether, at least twenty two people developed anthrax infections, eleven with the life-threatening inhalation type. Five died, including two mail workers at the US Postal Service Brentwood sorting facility in Washington. The investigation was massive and, to some, the identity of the actual perpetrator remains in doubt. The FBI announced that the culprit was not, as so many had assumed, some Islamic terrorist, but Bruce Ivins, a biodefense researcher at Fort Detrick who was reported to have mental health problems. He committed suicide in 2008 before he could be prosecuted. For many reasons, I'm convinced that he was the lone terrorist in this tragic story. I'm equally convinced that there are other Bruce Ivins–like scientists in labs around the world who could do it even "better" today.

The casualty count from this one episode was, fortunately, low, though even one death is too many. But it cost a total of more than a billion dollars to clean up and decontaminate the Hart Senate Office Building and other congressional and media offices and postal facilities exposed to the handful of letters. With work progressing around the clock, it took three months to reopen the Hart Building. It took more than two years to reopen the Brentwood facility and more than three years to reopen one in Hamilton, New Jersey.

The primary aim of terrorists, obviously, is to cause terror. And infectious agents historically have been the greatest source of terror in all of society, dating way back to before the Middle Ages.

Preparing for the naval battle against King Eumenes II of Pergamum in 184 BC, Hannibal directed his sailors to fill pots with "serpents of every kind" and hurl them onto the enemy ships. In 1346, the Tatar army, in its siege on the Black Sea port city of Caffa, catapulted dead plague victims over the fortified walls of the city, igniting an epidemic.

During the siege of Fort Pitt, Pennsylvania, during Pontiac's War in 1763, militia commander William Trent wrote that he had sent the Ottawa Indians "two Blankets and an Handkerchief out of the Small Pox Hospital," adding, "I hope it will have the desired effect." It probably did, likely triggering the "raging epidemic" that shortly followed. The suggestion had come from Field Marshal Jeffery Amherst, for whom the prestigious Massachusetts college is named.

In World War I, vials of anthrax were found in the luggage of captured German spy Baron Otto Karl von Rosen, intended to infect animals used by the Allies. In World War II, Japanese planes spread contaminated rice and plague-infected fleas over Zhejiang Province, China. During the Cold War, the Soviet Union and the United States maintained large germ warfare research programs. Prior to the end of apartheid, the repressive South African government maintained an arsenal of HIV, Ebola, and other lethal agents in case the regime was attacked.

President Nixon curtailed the American offensive bioprogram in 1969, when he concluded that the use of biological weapons could not achieve any legitimate military aims. From then on, the doctors, scientists, and technicians at Fort Detrick engaged only in biodefense research. But the Soviets kept right on working on a wide array of bioweapon development and production.

I'll never forget a meeting Mark Olshaker and I had in 1998 with Ken Alibek one Saturday morning in a coffee shop near his home in northern Virginia. We had been connected by a CIA contact. Alibek, who held an MD and a PhD in microbiology, was a friendly, heavily accented, though soft-spoken Kazakhstani immigrant with mildly Asian features. But before the fall of the Soviet Union, he was known by his original name, Kanatjan Alibekov, had held the rank of colonel in the Soviet army, and was deputy director of Biopreparat, the massive, secret, offensive biowarfare establishment, where he was responsible for developing the worst natural microbes into even worse weapons of war. He left Russia just after the fall of the Soviet Union, having become convinced that the United States really had abandoned offensive bioresearch and that his superiors were lying about the need to continue their deadly development. Though he insists that he didn't defect, he concedes that he left despite direct orders from the KGB not to leave the country.

As Mark and I sat there with Alibek and his wife, Lena, he calmly recalled his research and described the agents he had worked with: anthrax, brucellosis, glanders, Marburg, plague, Q fever, smallpox, and tularemia, among other toxic bugs. All of them had been bomb and missile ready. He said they had developed 2,000 strains of anthrax alone in an attempt to make it as deadly as possible.

Most frightening of all was Alibek's recounting of experiments to insert the gene of Venezuelan equine encephalitis, a mosquito-borne virus that attacks the brain—into vaccinia, the smallpox vaccine. If this were successful, it would be only a small step to put it into smallpox itself, with which he assured us Biopreparat had been well stocked, creating a superweapon against which the US vaccine would not work. This research was part of an organized program, he told us, known as the Chimera Project.

In spite of biological warfare's long history and our experience of it in my lifetime, in the more than a decade and a half since the 2001 anthrax attack, our state of unreadiness and denial has remained more or less the same. What has changed, though, is our gain-of-function capability. Tools to fundamentally alter how a virus or bacteria kills, or even potentially transmits, that did not exist in 2001 are now in the hands of many thousands of scientists in universities, colleges, high schools, and commercial labs and even in the possession of DIY amateurs—the ones tinkering in their garages and basements. We no longer can concern ourselves just with highly funded national and institutional defense labs. Information on how to gin up a potential killer microbe with new lab technology tools is readily available on the Internet.

Twenty years ago there were five class A agents of greatest concern for bioterrorism: anthrax, smallpox, plague, tularemia, and hemorrhagic fever viruses such as Ebola. Today, I worry primarily about anthrax, smallpox, and any microbe that we can change through our new hyper-lab tools to be readily transmissible to people or animals and resistant to current treatments or vaccines.

Anthrax—*Bacillus anthracis*—is a particularly effective bioweapon. It doesn't transmit person-to-person, but when dried out, the bacteria preserve themselves as tiny, virtually weightless spores that are hardy enough to last for decades or longer. Archeologists have even found evidence of them in Egyptian tombs. When those spores are inhaled and reach the moist, comfortable environment of the lungs and gastrointestinal tract, they germinate, reverting back to their active form and releasing three deadly protein toxins. Inhalation of anthrax in the lungs causes pneumonia that kills between 45 and 85 percent of untreated victims. In dried form, anthrax can be hidden in any white powder and will not arouse the suspicion of airport security workers or anyone else.

Back in 1993, the congressional Office of Technology Assessment produced a report entitled *Proliferation of Weapons of Mass Destruction: Assessing the Risk,* comparing the potential impact of chemical, biological, and nuclear weapons on Washington, DC. It concluded that a small airplane dispersing only 100 kilograms (about 220 pounds) of anthrax spores would kill more people than a Scud-class missile carrying a hydrogen bomb. The H-bomb would kill between 570,000 and 1.9 million in a 300-square-mile area, depending on such factors as weather and exactly where it was dropped. The anthrax dispersal would kill between 1 and 3 million under similar circumstances.

The late William "Bill" Patrick was a brilliant scientist and a friend to both Mark and me. He used to head up the American bioweapons program at Fort Detrick. Bill made a habit of carrying around a vial containing 7.5 grams of a harmless bacterial culture that looks just like anthrax under a microscope. In March 1999, testifying on Capitol Hill before the House Permanent Select Committee on Intelligence, he pulled out his vial, explained what it was, and declared, "I've been through all the major airports and security systems of the State Department, the Pentagon, even the CIA, and nobody has stopped me." Seven and a half grams, by the way, would be just about the exact amount needed to kill everyone in a structure the size of a Senate or House office building.

Anthrax can be treated with certain broad-spectrum antibiotics like ciprofloxacin (Cipro), but quick diagnosis is essential and treatment can take weeks or months. And experimental lab work has already proved how easy it would be to develop antibiotic-resistant strains.

Bioweapons are unlike any other of their brother weapons of mass destruction, and our response strategies for other WMDs will not work against them. As horrific as it is to think of two jetliners hitting and bringing down the World Trade Center towers, that was a readily "survivable" tragedy for New York City and the nation. At the end of the day on September 11, 2001, the terrorist act was over and the recovery could commence. With a bioterror event, the end of the day would be only the beginning, and no one would even know it yet. We likely wouldn't recognize it for a week, by which time the initial victims would have carried their deadly infection to all parts of the United States and much of the world.

Even with biologic agents that are not transmissible person-to-person, the challenge is daunting. The Mall of America in Bloomington, Minnesota, not far from where I live, is the largest shopping center in the United States, with an average of more than 100,000 visitors a day from all over the world. If anthrax were efficiently dispersed throughout the sprawling mall, there would easily be many thousands of cases and thousands of deaths as local healthcare systems were overwhelmed. The victims wouldn't even know they'd been targeted until several days had passed and fever, chills, chest pain, shortness of breath, fatigue, vomiting, and nausea set in. For many of them, recognition would come too late.

It would be an event of historic proportions that could never be forgotten, not only because of all the death and disease and the almost unimaginable panic that would ensue, but also because it would simply be too big and complex a task to decontaminate the entire mall complex. And you couldn't just tear it down, either. The AMI building in Florida was closed off for more than five years because of the risk of spreading the anthrax spores to the surrounding community. After a monumental cleanup effort it was finally declared anthrax-free in 2007. A contaminated Mall of America—many times larger than the AMI building—would just sit there as an abandoned, hulking mass on the Minnesota prairie—as toxic and uninhabitable as Chernobyl.

Number two on my list of the big three is smallpox. Despite the fact that it has harmed no one in almost forty years, smallpox remains one of the scariest monsters on earth. Its toll on human history tops a billion deaths, and an even greater residual effect of acute suffering and disfigurement. So powerful has been its cultural influence that it is perhaps the only disease that is represented by gods and deities in various cultures. Today we no longer attribute the virus to gods, but the mere thought of its reemergence haunts the fevered dreams of all responsible public health officers.

In the late 1990s, we were vulnerable. We had no way of protecting the world population against an accidental or intentional release. Vaccine stockpiles were almost nonexistent—there had been no need for them for so long—and of what was still around, we had not assessed the remaining potency.

In 2014, vials marked "Variola" were discovered in an unused portion of a storage room in an FDA lab on the NIH campus in Bethesda, Maryland. Apparently, the vials dated from the 1950s and no one noticed them when the lab was transferred from NIH to FDA control in 1972. Now, what if the discovery had been made by a disgruntled lab employee of the type we mentioned earlier? I think the implications are clear. And I believe it very likely that other smallpox samples are stored away in some researcher's freezer, waiting to be discovered someday.

But here's where the story gets more complicated, and even scarier.

The twenty-first century, as we've all seen, has witnessed an explosion of advances in the genetic sciences. Decades after Watson and Crick figured out the double-helical structure of the DNA molecule, we are now able to explore the arrangement of the thousands of molecules of adenine, thymine, cytosine, and guanine that make up each plant's and animal's genetic code. In the shadow of the monumental, government-funded Human Genome Project, gene sequencing of various organisms became a reality.

In 2002, with support from the Defense Department's Defense Advanced Research Projects Agency, the same organization that developed the Internet, Dr. Eckard Wimmer, distinguished professor of molecular genetics and microbiology, led a team at Stony Brook University on Long Island and synthesized the poliovirus from scratch. The virus contains 7,500 base pairs of genetic information, the critical combinations of adenine, thymine, cytosine, and guanine that constitute the code of life. Making a disease-causing poliovirus from scratch would have been considered science fiction just years before. This was an astounding and historic scientific event: the first creation of a disease-causing virus from off-the-shelf genetic material, simply following the instructions of its published sequence.

With only 7,500 base pairs, polio is a relatively simple virus, compared to smallpox. HIV has 10,000. Back in 1994, J. Craig Venter and his colleagues determined the entire genetic code for the smallpox virus: a whopping 186,102 base pairs. If polio represented a 100-story genetic building, smallpox was a 1,600-story structure, so we didn't need to worry much about anyone building it in his or her laboratory. No one could do with smallpox what Wimmer did with poliovirus.

But as the technology raced forward, the superskyscrapers of genetic engineering became more and more attainable. Today, it will soon be possible, if it is not already, to re-create the smallpox virus in a lab just as Wimmer re-created polio. In fact, in an October 2014 opinion piece in the *New York Times* entitled "Resurrecting Smallpox? Easier Than You Think," a highly respected professor at the University of Southern California, Leonard Adleman, described how his lab or others might make smallpox virus using a similar approach. In other words, we can now build 1,600-story genetic buildings.

Will it be easy? Certainly not. But it will be far simpler than building and detonating a nuclear device, and that is something we worry about all the time. What is more, through gain-of-function techniques, terrorist-employed scientists might be able to modify or enhance their new variola virus so that we are not protected by our current vaccine.

To be effective, a weapon must possess certain attributes. It must be within the economic means and scientific expertise of the prospective deployer. It must be capable of reaching its intended target. It must be able to limit collateral damage to those not intended as targets. And its use must result in the desired outcome.

Few other weapons meet this criteria so well for terrorist use, considering they are not expensive compared to other WMDs, reaching a target is easy, terrorists don't consider anyone collateral damage, and the desired outcome—panic and residual fear—are assured. Though the twelve fatalities from the 1995 sarin gas release in the Tokyo subway system were far fewer than the Aum Shinrikyo cult leaders had hoped, they certainly achieved their goal of creating fear and social disruption.

Moreover, the delay between release/infection and the onset of symptoms compounds and prolongs the terror and makes it that much more difficult to track, identify, and apprehend the terrorist.

Smallpox meets all of these criteria. What we don't know is how many designer bugs of the near future may also measure up. If we have no effective response, then terrorists of any stripe may be able to actually achieve their goals. For the first time in human history, a few evil individuals may have the power to upset the political balance, security, health, and economic well-being of the entire planet.

What kinds of individuals are we talking about? In the current world climate, Islamic extremists, either working alone or under the aegis of a group like ISIS, have to be at the top of the list. But they are far from the only ones. We also have to consider scientists who are mentally ill or willing to sell their knowledge and services to the highest bidder. Many countries, including the United States, have long histories of domestic terrorism, in our case extending from the Ku Klux Klan to Timothy McVeigh.

There are any number of reasons their twisted minds might come up with for wanting to kill their fellow citizens in this most insidious way. And we have seen cases over the years of lab employees who consider themselves underappreciated and better than their positions and want to prove it in this sick way.

This hardly exhausts the possibilities. Unabomber Theodore Kaczynski, a man with a near-genius IQ, railed against the soullessness of industrialized society from a lonely cabin in Montana. Kaczynski knew how to make bombs. If he had received his PhD in biochemistry instead of mathematics, he might have gone the bioterror route. As Mark and his longtime writing collaborator, former FBI special agent John Douglas, have shown in their books, many of these pathologically antisocial types wage a constant internal war between deep-seated feelings of inadequacy and equally powerful notions of grandiosity and entitlement, together with resentment that they are ignored by the rest of the world.

To understand how unprepared we would be in the event of a bioterrorist release of smallpox, let's look at an actual case involving a closely related but less serious disease.

In 2003, a single ten-year-old female patient was admitted to SwedishAmerican Hospital in Rockford, Illinois, suffering from a case of monkeypox. You probably haven't heard much about this disease, because it comes from the same orthopoxvirus family as smallpox, and the smallpox vaccine confers immunity in humans, so it was never much of a worry. But both viruses can cause similar devastating symptoms. And while monkeypox has a much lower overall fatality rate, though still high at about 10 percent, it has one characteristic that smallpox does not have: It can be transmitted across species.

Isolated in African monkeys in the 1950s (hence the name), monkeypox can thrive in squirrels, mice, and a host of small rodents in parts of Central Africa. The young patient, Rebecca, caught it from a pet prairie dog purchased at a pet store that had some Gambian pouched rats as exotic pets. These rats had been shipped from Ghana to Texas and from there to the store in suburban Chicago. That's how easily infectious diseases can hitch rides around the world.

Rebecca was one of thirty-seven confirmed monkeypox victims of the US outbreak that summer and the only one at SwedishAmerican Hospital. Nonetheless, when she developed pox pustules over her entire body—even inside her mouth and throat, a high fever, pain, and trouble swallowing, chaos and panic erupted through the entire hospital. Few on the medical or nursing staff had been vaccinated for smallpox, recently or at all, and there were even practical and ethical arguments on whether to admit her or try to transfer her to another hospital. Some staff members literally feared for their lives and others refused to take prophylactic smallpox vaccinations because of the side effects.

There is no cure for monkeypox. Rebecca was put into isolation and anyone authorized to get near her had to wear a respirator and full protective isolation gear. No one was allowed to touch her skin without protection.

Thankfully, she recovered with only some residual scarring to show for her ordeal. But if treating this one small patient almost undid the hospital staff and created lasting emotional wounds, imagine if this had been smallpox, and it wasn't limited to a single patient.

In the event of a smallpox attack, victims won't even know they've been attacked for at least a week, and neither will anyone else. By that time the perpetrators will be long gone. Before long, some of those affected will start showing up in doctors' offices and hospital emergency rooms with the cliché vague, flu-like symptoms, including headaches, backaches, high fever, and possibly nausea and vomiting. Most will be sent home and told to drink fluids and get plenty of rest. Some will feel bad enough that they'll be tested for serious problems like meningitis, but the results will be negative. A few of the more perceptive physicians will consider a staph infection, possibly foodborne, but that diagnosis will not pan out.

When these same people come back in with spreading rashes, the doctors will start thinking more exotically, but the patients won't respond to any of the pharmacopoeia of antibiotics they throw at the bug. The body nodules will turn to hard pustules and then start erupting and oozing. By that point, doctors will have stopped scratching their heads and will be whispering to themselves or their colleagues about what they see happening but can hardly believe. None of them will ever have seen an actual case of smallpox before.

At that point, all hell will break loose. Every frontline physician and public health official will be on the phone with state health departments, the CDC, or anyone they can think of. It will quickly become clear to the CDC's and HHS's emergency coordinators, who will be reporting to the White House on an hourly basis, that there are clusters of sick patients all over the country, with a larger number concentrated in the New York, New Jersey, Pennsylvania, and Connecticut region. Absenteeism will be found to be higher than normal for this time of year.

The White House will call in whoever might have any information, including the living legends of smallpox eradication they can find. They will order the release of the entire strategic reserve of smallpox vaccine developed under the leadership of Secretary of Health and Human Services Tommy G. Thompson after 9/11. First responders and frontline treatment personnel will be first to receive the vaccine, as well as military troops and law enforcement officers. The first approach will be to try the ring vaccination strategy that Bill Foege devised for India in the 1970s, but as the number of cases increases, this may not be feasible. Meanwhile, the death figures will start to come in and national panic will set in. Everyone will be desperate to get their hands on the vaccine. Drugstores will be looted, though they have no vaccine, and several governors will call the National Guard. A black market for the vaccine will quickly form. The president will urge calm, saying that everyone will get the vaccine eventually, but when reporters try to pin him or her down on a timetable, the response will be that it is too early to be specific.

At a White House meeting, hastily drawn quarantine plans will be described to try to get ahead of the spread. Mass quarantine will not have been attempted for more than a hundred years and the attorney general won't even be sure who may order them. But the directors of the CDC will say it may be a moot point; there will be too many clusters to try to quarantine large populations, particularly now that reports are coming in daily of new cases in Europe, Asia, Africa, and South America. All will have been visitors to the United States three weeks before. Other nations will continue their demands for a UN-sponsored quarantine of the United States, Canada, and Mexico.

The death rate will continue to rise. Funeral homes will refuse to accept bodies. Hospitals, with no alternative, will store them in large refrigerated trucks. The media will run features on the New World Indians at the time of Columbus, devastated by smallpox and other diseases for which they had no herd immunity. The stock market will fall by 75 percent.

We could go on and on with this scenario, and there is no telling how many generations of spread we would go through before we finally had the crisis under control. Suffice it to say that it would overshadow 9/11 many times over and leave a permanent scar on the American and world psyche.

To paint an even grimmer picture, there would be nothing to prevent the terrorists from "reloading" and staging another release just as we began to recover from the first one. And the ultimate horror: What if scientists working for the terrorists figured out a way to alter the smallpox genome, to make it so that immunity conferred by our existing smallpox vaccine isn't protective?

In October 2015, a nonpartisan blue-ribbon panel of experts cochaired by former senator Joseph Lieberman of Connecticut and former Pennsylvania governor and the first secretary of Homeland Security Thomas Ridge produced a report entitled *A National Blueprint for Biodefense: Leadership and Major Reform*

Needed to Optimize Efforts. The subtitle is actually a rather mild description of the panel's findings.

The basic and repeated message of the report is: "The United States is underprepared for biological threats." Despite the US Commission on National Security/21st Century, the National Commission on Terrorist Attacks upon the United States, the Commission on the Intelligence Capabilities of the United States Regarding Weapons of Mass Destruction, and the Commission on the Prevention of Weapons of Mass Destruction Proliferation and Terrorism, the report concludes, "the insufficiency of our myriad and fragmented biodefense activities persists because biodefense lacks focused leadership."

It gets worse: "Simply put, the Nation does not afford the biological threat the same level of attention as it does other threats: There is no centralized leader for biodefense. There is no comprehensive national strategic plan for biodefense. There is no all-inclusive dedicated budget for biodefense."

I agree. One of the scariest and most interesting elements of the report is a fictional address by the chairman of a Senate–House of Representatives "Joint Inquiry into Administration and Congressional Actions Before and After the Bioterrorist Attacks of 2016" (a hypothetical attack projected into the then future). In the hypothetical scenario, the aerosol release of a genetically modified Nipah virus (an agent causing encephalitis and respiratory distress first identified in 1998 in Malaysia) in Washington, DC, led to the deaths of 6,053 individuals, including senators, representatives, and staff members, and the sickness and incapacity of tens of thousands more. A coordinated release targeted livestock in rural communities.

The fictional chairman's statement neatly summarizes our very real current shortcomings:

> The terrorists were successful because the government—including Congress—failed. They took advantage of our fail-

ure to achieve early environmental detection of the agent, failure to quickly recognize its occurrence in livestock, failure to rapidly diagnose the disease caused in sick patients, failure to consistently fund public health and health care preparedness, failure to establish sufficient medical countermeasure stockpiles, failure to make sure that non-traditional partners communicate. Ultimately, they took advantage of our failure to make biodefense a top national priority.

Sadly, much as the 9/11 Commission observed in its analysis of the attacks of 2001, the attacks of 2016 occurred because of another "failure of imagination."

Failure is the report's central theme. To the failures of prediction, early warning, and detection, the chairman says, "we must now add the failure to appreciate the threat, generate political will, and take action in the face of looming danger." That, in a nutshell is where we are.

What can we do?

Bill Gates realizes the enormity of the challenge, even with his resources. "If you can tell me how to write checks and stop bioterrorism, then sure," he said to us. "I'm a risk-adjusted kind of guy; I'll write checks. But who are you writing the check to? What is it you're doing?" When we're talking about this kind of event, he rightly concludes, "This is governmental stuff."

And yes, preparedness would require a lot of money, but money is not enough; it would also require organization and robust planning. We cannot content ourselves to be merely reactive to these threats. If a bioterror event does occur, we need to have a public health and medical care system already in place to meet the immediate challenges of a situation that is no longer unthinkable.

Some in public health and medicine are openly critical of spending even limited government resources on preparing for something that "might happen" when every day Mother Nature is throwing at us real and serious infectious disease challenges. Their point is well taken. But we need to remember that for a long time, the intelligence community doubted that terrorists had the organization and resources to launch a large-scale attack in the United States. September 11, 2001, certainly disabused all of us of that assumption. And what many analysts still fail to recognize is that a bioattack can be carried out on a very small scale and still have major destructive effects.

Among the action items recommended by the blue-ribbon panel is the creation of a national intelligence manager for biological threats, who would coordinate all efforts and understand the critical importance of the One Health concept, since animals could be affected as well as humans, and since 60 percent of emerging infectious diseases are leaping into the human population through animals.

I also agree with the report's recognition that biodefense has to be addressed at the state and local levels, since any attack will first be the domain of emergency first responders and hospital emergency room personnel, who will have to recognize what they are seeing. The report recommends that hospital accreditation and federal funding levels and reimbursements should be contingent on preparedness to handle unexpected biological events. To be effective, federal funding to the states will have to be significant to allow them to meet and be prepared for the challenge.

Among the panel's strongest recommendations is the coordination of communication and resources between NIAID and BARDA, both of which have an important role to play in biodefense. The problem here is similar to that of vaccine development in general. A fair amount of money is put into basic research and early-stage development for what we call MCM—medical countermeasures—but relatively little goes toward the actual production and distribution of treatment agents. The specific recommendations in this realm are: (1) ensure that

NIH research supports civilian countermeasure priorities; (2) ensure that funding allocations are appropriate to meet the need; and (3) require a biodefense spending plan from NIAID. In real life, however, "administrations have touted the success of the [emergency preparedness] program while simultaneously scaling back their budget requests."

As we've shown through our anthrax examples, even after the horror of a bioterrorist event has run its course, environmental remediation will remain a formidable, if not insurmountable, task. The plain fact is that we really don't know how to do this, and additional research is urgently needed. And again, while the Environmental Protection Agency has some responsibility, there is no clear-cut law, set of rules, or even any official guidance on how to go about such a task.

I don't agree with everything in the blue-ribbon report. There are too many vague and squishy recommendations that begin with words like "empower," "enable," "require," "develop," "incentivize," "assess," "determine," and "align." But the questions the report poses and the overall message of a fundamental lack of preparedness for one of the direst conceivable threats to the well-being of our nation and others around the world must not be ignored.

According to the myth, after all of the horrors have flown out of her opened box, Pandora notices that it is not quite empty.

> At the bottom of the box was a quivering thing. Its body was small; its wings were frail; but there was a radiance about it. Somehow Pandora knew what it was, and she took it up, touched it carefully, and showed it to Epimetheus. "It is Hope," she said. "Do you think it will live?" asked Epimetheus.

Having opened Pandora's box, it is up to world leaders, and all of us, to give that last bit of its contents every possible chance.

CHAPTER 12

Ebola: Out of Africa

The future is already here — it's just not very evenly distributed.

— William Gibson

Why were we surprised in 2014?

Ebola was first identified in 1976 in near simultaneous outbreaks in Nzara, South Sudan, and Yambuku, Zaire, now the Democratic Republic of the Congo. Like its predecessor Marburg, Ebola is a filovirus, so named because of the looping, filament-like morphology of its virion. The disease name was derived from the Ebola River, near the village where the Congo outbreak occurred. From 1976 to 2013, there were twenty-four documented outbreaks in Africa, the largest occurring in 2000. One was in Gulu, Uganda, which numbered 425 cases and claimed 224 lives. The toll for most of the other outbreaks had been considerably smaller. And that is what most scientists and public health officials expected to continue to see, rather than a full-blown epidemic in 2014.

Ebola lives mysteriously, deep in the equatorial forests of Central Africa. To this day we aren't certain of the animal reservoir, though we think it is probably fruit bats. Anytime it emerged in the human population, the spread was in areas so remote and isolated that most cases could be managed with limited resources and small public health support teams.

The greatest risk of human-to-human transmission occurred in medical clinics and hospitals where cases were brought for care. Without modern infection practices, including gloves and other worker protection equipment, these healthcare facilities often become the "case magnifiers." The first response to stopping an emerging Ebola outbreak was to bring in infection control experts and the necessary infection control–related medical supplies to stop transmission in that setting. Despite no effective specific Ebola treatment or vaccine, these standard approaches worked, and the disease died out relatively quickly.

Then, in March 2014, Ebola showed up not in its expected lair of equatorial Africa, but in forested areas of southeastern Guinea, on the west-central coast. It has been postulated that the first case, which ignited the subsequent West African outbreak, was a toddler who may have caught the virus from contact with bats in a hollow tree near his tiny village. Two days after the onset of his symptoms—high fever, vomiting, and bloody diarrhea—he died.

Many factors contributed to the spread of the 2014–15 epidemic, including adherence to traditional funerary and burial practices that involved extensive physical contact with infected dead bodies; greatly amplified transmission in the crowded slums of Monrovia, Freetown, and Conakry; structural inadequacies in local healthcare when Ebola cases could not be separated from other patients who did not have Ebola and which therefore, in the WHO's words, "ignited multiple chains of transmission"; and lack of equipment or trained personnel to provide appropriate care. A significant number of people hid their ailing relatives rather than surrendering them to hospitals or clinics that could do nothing for them and where they would die alone. Unprotected African physicians and nursing staff were stricken and died in unconscionably high numbers. The failure of the WHO and other international groups to recognize the problem and their inability to act also prolonged the crisis.

As WHO director-general Dr. Margaret Chan observed at a conference in London in September 2015, "A disease like Ebola will expose every gap in health system capacity and exploit every opportunity opened by these gaps." This has always been true.

So what was different this time?

The short answer, which I described in a July 2014 *Washington Post* op-ed, is this: Ebola virus didn't change. Africa changed. This simple fact had infinitely complex implications for this outbreak and will have for any yet to come.

First, the deforestation in Guinea from large-scale foreign mining and timber operations made it easier for Ebola to escape from animal populations deep in the forest. Second, the residents of Guinea, Liberia, and Sierra Leone travel much farther and have many more interpersonal contacts than they did in previous decades. Contact tracing—following up on all the contacts with the infected person—is much easier if those contacts live within a short distance of the case rather than far away and spread out.

With modern transportation, family members may travel hundreds of miles to be with sick loved ones. And much more of the West African outbreak area is urbanized than were the locations of many of the previous outbreaks, leading to faster and denser spread, particularly in the slums of the three capital cities. All these factors made this virus hyperevolutionary. In the first four months of the outbreak, there were more human-to-human transmissions than most likely occurred in the last 500 to 1,000 years. That represents a lot of throws of the genetic dice.

The virus replicates efficiently in various cells throughout the body, causing an extreme inflammatory response and septic shock. While the commonly evoked symptoms of blood dripping from eyeballs and internal organs turning to mush are more sensationalistic than clinically accurate, the actual disease is gruesome enough. It can start out with fever, chills, a severe headache, joint and muscle pain, and fatigue about five to ten days after exposure, and progress to nausea and vomiting, bloody diarrhea, rash, gut pain, bruising, and bleeding. In the end stages, blood actually can ooze from the eyes and mouth and rectal bleeding is not uncommon. Even more devastating is the internal bleeding that collects in the spaces between organs, caused by decreased clotting. In fatal cases, death is often caused by low blood pressure leading to circulatory failure and severe fluid loss.

Because of the quick, horrible symptoms, frequently leading to an equally horrible death, Ebola struck a chord of fear in a way that many other more common and prevalent infectious diseases have not. The 2014–15 outbreak resulted in more than 28,600 cases and 11,325 deaths and left more than 30,000 West African children orphaned.

And because of its rarity, Ebola hadn't been factored into individual threat matrices as malaria, TB, AIDS, and vaccine-preventable and diarrheal diseases had. We saw this not only in west-central Africa, but here in the United States, where numerous people were afraid of contact with anyone who had been anywhere on the African continent in recent weeks. "Out of an abundance of caution..." was a commonly heard refrain from political figures and even some public health officials.

In actuality, those individuals were in virtually no danger. To date, the primary transmission route for Ebola is from the bodily fluids of an infected person. Unlike HIV, which is transmitted through sexual relations, exposure to infected blood through a wound, transfusion with HIV-contaminated blood, or birth to an HIV-infected mother, Ebola can spread through simply touching an infected person or their body fluids and possibly by breathing in aerosolized bodily fluids caused by certain medical procedures. Two of the most common methods of transmission in the epidemic: handling of dead bodies in funerary practices and caring for stricken patients in hospitals or at home. But unlike influenza, which is contagious in infected persons even before they are ill, Ebola victims are not contagious until they actually begin showing symptoms. And those symptoms, as we have noted, are hard to miss.

Fear took over from rational response on many levels. Certain Pentecostal church leaders in Africa tried to deny Ebola's existence, then claimed it was divine punishment for promiscuity and homosexuality. There were other examples of cultural beliefs trumping science. In Monrovia, people brought their sick relatives to be healed in church, and as many as forty pastors died after contracting the virus from ministering to their afflicted congregants.

In September 2014, while giving a breakfast briefing to senators and congressmen in a Senate conference room in the US Capitol, I had a passionate exchange with one senior congressman who said he wanted to introduce legislation banning all flights to and from affected areas of Africa into the United States until the epidemic was certified as over. I pointed out to him that if physicians, nurses, and other public health workers knew that if they contracted the virus while treating patients they would not be returned to the United States to receive medical care, we would have a sudden shortage of people willing to go over and fight the outbreak, making it all the more likely it would spread over here. I asked him how he proposed to get supplies into the hot zone without airplane flights. Fortunately he concluded that maybe the flight ban was not the best way to deal with the situation.

Other legislators and some governors called for extended quarantines on all healthcare workers returning from the field—another "abundance of caution" initiative. Many in the public health community unflatteringly referred to New York governor Andrew Cuomo and New Jersey governor Chris Christie as "Drs." Cuomo and Christie after their scientifically erroneous and misguided statements about the public health reasons for quarantining healthcare workers after their return from working with Ebola patients.

I found myself somewhere between proponents of two extreme positions: those calling for a twenty-one-day isolation-room quarantine for anyone who had even a brief conversation with an Ebola patient, even across a large room, and those claiming that any follow-up of a healthcare worker who cared for an Ebola patient was an intrusion into their personal rights and without any medical or public health justification.

Every piece of science-based information we had at the time supported the proposition that persons infected with Ebola were not going to transmit the virus to anyone in the first day or two after the onset of clinical symptoms. And there are two reasons why potentially exposed healthcare workers had every reason to report such symptoms immediately. First, they had willingly laid their lives on the line to care for Ebola patients. Who would believe they would then put others in harm's way if they had reason to think they might transmit Ebola virus?

But, even if we are skeptical of such an altruistic point of view, healthcare workers were well aware that when early intensive medical treatment is initiated with the Ebola infection, the survival rate improves dramatically. So what healthcare worker, having just cared for an Ebola patient, is going to be hunkered down at home or strolling the streets if they might have early symptoms of Ebola?

This was true for the three American healthcare workers who sought medical care when they became symptomatic. None of them transmitted the infection while out and about. Since in virtually every case, a person is not contagious with Ebola until he or she has notable symptoms, healthcare professionals' self-monitoring will prevent Ebola virus transmission to family members, colleagues, or strangers on the street or in a subway car.

On the other hand, I did find troublesome the attitude of a very few returning healthcare workers who maintained that public health or the long arm of government had no right to intrude into their personal lives. It only reinforced in the minds of many in the public and some politicians the idea that we, the medical and public health communities, were putting ourselves first and didn't worry about transmitting the Ebola virus to others. Unfortunately, we did a poor job explaining to the public that self-monitoring by healthcare workers returning from Africa or those who cared for Ebola cases hospitalized in the United States would protect everybody.

Will the Ebola virus always be transmitted in the ways outlined here? The one previous time that the Ebola virus had appeared in America was in a holding building for laboratory-bound crab-eating macaque monkeys in Reston, Virginia, in 1989. This outbreak was the backdrop for Richard Preston's 1995 bestseller, *The Hot Zone.* Though all the monkeys died of the disease or were euthanized to prevent spread, the Reston strain—which is different from the one that caused the outbreak in West Africa—turned out not to be infectious to humans. Unfortunately, that has not been the case for the four other known strains.

In Reston, the humans lucked out. But the affected monkeys were all in cages without touch contact with one another. So it is likely that this strain of the virus was transmitted via the respiratory route. Does this mean that the Reston Ebola virus ever could be transmitted via the airborne route to infect people? We don't know. Recently a group of researchers at the University of Kent demonstrated that not much genetic change would be needed in the viral genome of the Ebola virus for it to adapt to novel hosts—like Reston Ebola being able to infect humans. They concluded, "Human pathogenic Reston viruses may emerge. This is of concern since Reston viruses circulate in domestic pigs and can infect humans, possibly via airborne transmission."

In 2012, a team of Canadian researchers showed that Zaire Ebola, the same strain that caused the virus in west-central

Africa, could be transmitted via the respiratory route from pigs to monkeys, both of whose lungs are very similar to those of humans. If airborne transmission of Ebola virus to and by humans were to occur, it would be a game changer. That is a very, very big deal.

Though I was criticized as being alarmist for bringing this up in a *New York Times* op-ed I wrote in September 2014, I considered it—and still consider it—a possibility we cannot and should not dismiss. Prior to writing that piece, I had a number of conversations with some of the leading internationally recognized Ebola virologists and epidemiologists who were asking the same question privately, noting that the virus had passed through more humans in a few weeks than it had done cumulatively for decades before, and that this hyperevolution could favor a respiratory-transmitted virus. But they were reluctant to talk about the possibility for fear of being labeled scaremongers.

In March 2015, nineteen other authors and I, including some of these same leading Ebola experts from the United States, Europe, and Africa, published a comprehensive review of what we knew and what we did not know about the transmission of Ebola virus in *mBio,* a prominent microbiology journal. We stated, "Despite the lack of supportive epidemiological data, a key additional question to ask is whether primary pulmonary infections and respiratory transmission of Ebola viruses could be a potential scenario for the future. A fair amount of evidence suggests that such transmission could be possible, even without dramatic evolution or genetic changes in Ebola viruses (although viral evolution over time could enhance that possibility)."

One well-known Columbia University virologist, Dr. Vincent Racaniello, wrote the following about my *New York Times* op-ed on Ebola on his widely read blog shortly after its publication: "We have been studying viruses for over 100 years, and we've never seen a human virus change the way it is transmitted.... There is no reason to believe that Ebola virus is any different from any of the viruses that infect humans and have not changed the way that they are spread."

This statement is simply not true; we do have examples of viruses changing how they are transmitted. We need look no further than the Zika virus epidemic in the Americas. In February 2016, Racaniello wrote about Zika transmission: "Can Zika virus be sexually transmitted? Perhaps in very rare cases, but the main mode of transmission is certainly via mosquitoes."

Dr. Racaniello may now want to reconsider his blog post on the rarity of sexual transmission of Zika virus. By early summer 2016, we had confirmed that human sexual transmission of Zika, a vector-borne disease, is not rare and represents a newly recognized and important means of virus transmission. Many leading experts on mosquito-transmitted diseases have concluded that a mutation in the Zika virus has fundamentally changed the ways and magnitude of how it is transmitted in humans.

We can't write off the possibility that someday, airborne transmission of Ebola virus will occur in the community setting. I pray that never happens, and we don't have any evidence to date that it has occurred in West Africa. *Yet!* But when the science community shuts its collective mind to what Mother Nature might do because it's just too scary to contemplate, as some have done with Ebola virus transmission, we surely won't be better prepared for the next biologic curveball, whatever it happens to be.

As an example of how much we have yet to learn, we had always assumed that if a patient had recovered from Ebola, he or she was immune to the disease and not contagious to others. Ian Crozier is an American physician who was one of the heroes of the response in Sierra Leone. In May 2015 it was revealed that after his treatment and apparent complete recovery from Ebola, he was still harboring the virus in his eye. Subsequent studies have found the virus lurking in the testicles of some of the men who have recovered, adding to the fear of sexual transmission.

We have learned the hard way that these long-term infections can make extinguishing large Ebola outbreaks very challenging. As of May 2016, a series of flare-up Ebola outbreaks had occurred in West Africa, long after the outbreaks in each country had been declared over. In each instance, the recurrence likely started when a recovered Ebola case had sex with a previously uninfected person, or breastfed a child. Testing of semen and breast milk from recovered cases identified as the source of these new flare-ups shows that the Ebola virus may be present in these body fluids for many months and that these individuals still can transmit the virus. During that time, a small number of patients continue to have some symptoms, while the great majority have been asymptomatic.

Any one of these flare-up cases could ignite the next huge epidemic somewhere else in Africa. The big lesson we don't seem to have learned from 2014–15 is that this epidemic was not a one-off and that the task of putting out the obvious forest fire is not complete as long as the embers are still smoldering and giving off sparks.

The great fear all along was that the epidemic would spread beyond the three coastal nations. The first case in Nigeria has been cited as an example of how good surveillance and quick medical management averted a crisis in one of Africa's largest and most urbanized economies. But let us be clear: Taking nothing away from the admirable work of that nation's healthcare responders and the Federal Ministry of Health, in Nigeria we were a good deal luckier than we were effective.

First, when patient zero—Patrick Sawyer, a Liberian American lawyer who lived in Minnesota and who consulted for the Liberian government—arrived in Nigeria on July 20, 2014, from Liberia by way of Togo, he was already sick, so obviously so that he collapsed at Murtala Muhammed International Airport in

Lagos. He was taken to a hospital, where it still took three days to come up with a diagnosis.

As it happened, the public hospitals were on strike, so the patient was diverted to a private hospital called First Consultants Medical Center, which was better equipped to deal with an infectious patient. Even so, he infected nine healthcare workers before his Ebola was confirmed. One of the most important players in this episode was Dr. Ameyo Adadevoh, the hospital's chief medical officer. She treated Sawyer herself, placed him in quarantine, and kept him there, essentially against his will, resisting all efforts from the government and the institution itself to let him leave First Consultants. This, it was believed, would rid Nigeria of its problem. In fact, it would have done the opposite.

On July 28, Adadevoh started feeling symptoms. On August 19, she died. Today, she is regarded as a national heroine of Nigeria—a symbol of strength, commitment, and compassion.

Aside from committed healthcare workers like Adadevoh and her colleagues, what really helped save the day in Nigeria was the presence of polio eradicators, and we must give a lot of credit to Dr. Frank Mahoney of the CDC, who led the polio program in Nigeria and pulled his workers off to deal with Ebola. The CDC team provided command structure and Mahoney made sure its members worked closely with Nigerian health authorities in stamping out the disease in country.

And here we get into a series of terrifying what-ifs.

What if the polio eradicators had not been there? What if patient zero had not collapsed at the airport and had gone into one of the neighborhoods of Lagos? Two-thirds of Lagos's 15 million residents live in slum conditions without reliable, clean drinking water, electricity, or sewage disposal. If Ebola had taken root there, what took place in the three coastal nations would have become merely a disaster sideshow.

Lagos wouldn't have been the end of it. This type of megacity situation exists throughout sub-Saharan Africa. More people live in the slums of Kinshasa, in the Democratic Republic of the Congo, than in the three capital cities of Guinea, Sierra Leone, and Liberia combined. While Kinshasa is the largest city in the Democratic Republic of the Congo, with almost 14 million people, four other cities in the country have more than a million residents. In Nigeria, there are five other cities in addition to Lagos that number more than a million in population. Accra, Ghana, has more than 2.8 million residents. Any one of these cities is a gas tanker waiting to blow if the Ebola match is introduced.

What if we have to fight an Ebola war on multiple African fronts? Each year, thousands of young West African men and boys are part of a migratory work population not too dissimilar from US migrant farm workers. Crop-friendly rains wash over West Africa from May to October, defining the growing season. These young men typically help with harvesting in their home villages from August to early October. Afterward, they head off for temporary jobs in artisanal gold mines in Burkina Faso, Mali, Niger, and Ghana; cocoa nut and palm oil plantations in Ghana and the Ivory Coast; palm date harvesting and fishing in Mauritania and Senegal; and illicit charcoal production in all of these countries.

Like their ancestors before them, they use little-known routes and layovers through forests to avoid frontier checkpoints. They usually have Economic Community of West African States (ECOWAS) ID cards, providing free passage to and through all of the member states. It takes one to three days to travel from the coastal countries to these work destinations. There is no need for Ebola to hop a ride on an airplane to move across Africa. It can travel by foot.

"The fact that the Ebola epidemic was allowed to get as bad as it did is a frightening warning signal of what could happen," says Ron Klain. In mid-October 2014, he got a call from President

Obama, asking him to take over as the nation's Ebola czar for the duration of the crisis. Ron was not an obvious choice, having no medical background and, as he says, not even being qualified to administer a vaccine. A Harvard Law graduate, he went on to become chief of staff to Vice Presidents Al Gore and Joe Biden. President Obama's choice of Klain was widely criticized, but it turned out to be an inspired one. Klain is an expert in the rapid formation of government policy in the face of crisis, and in the coordination of a complex, interagency response to the crisis, and that is exactly what we needed.

"Yes, in the end, the death toll from Ebola was a small fraction of the worst-case forecasts from the CDC," Klain concludes, "and there can be little dispute that many thousands of lives were spared." And while the peoples of the affected nations made the difficult cultural and behavioral changes to slow the spread of the disease and took courageous steps to care for their families and neighbors, this locally powered effort was substantially aided by an unprecedented global response—a response led by the United States and a number of other countries as well as nonprofit organizations such as Doctors Without Borders.

But despite the experience of seeing the United States successfully mobilizing a total of more than 30,000 government workers, contractors, military service members, and volunteers in various areas of the response, Klain says, "A future epidemic could pose a far more challenging scenario."

And it's not just the emerging world that isn't prepared. As Klain noted, "There is not a city in the U.S. with more than three isolation beds, other than New York. New York has eight."

Yet no coordinated international response plan exists.

There is only one rational and comprehensive way to protect ourselves from another, and possibly much larger, Ebola epidemic: to develop, manufacture, and deliver an effective vaccine.

But as Dr. Seth Berkley, CEO of Gavi: The Vaccine Alliance, pointed out at a TED Talk while the virus was still raging, "The people most at risk from these diseases are also the ones least able to pay for vaccines. This leaves little in the way of market incentives for manufacturers to develop vaccines, unless there are large numbers of people at risk in wealthy countries. It is simply too commercially risky."

Despite this, the global community has made some strides in Ebola vaccine development since the outbreak began in West Africa in 2014. Thirteen potential vaccine candidates were tested in Phase I and/or Phase II clinical trials. In addition, three Phase III efficacy trials were started in Africa: one each in Guinea, Liberia, and Sierra Leone. One vaccine, known as the recombinant vesicular stomatitis vaccine (rVSV-ZEBOV) and made by NewLink Genetics and Merck, demonstrated preliminary evidence of protection.

With the outbreak in serious retreat and progress being made with vaccine work, many in the international community concluded that the Ebola crisis in Africa was over and never to be heard of again. Realistically, that's not at all the case. Without ongoing commitment from the global public health community, progress toward getting vaccines for Ebola approved could falter as memories of the outbreak in West Africa fade. Near the beginning of the 2016 Zika outbreak, American lawmakers decided to take the remaining Ebola funds to fight Zika, thereby giving neither disease the attention it warrants.

As of August 2016, a number of vaccines had advanced to various points along the clinical trial pathway. But no vaccine had yet been approved by regulators, and until one or more of them is approved and ready to stockpile in anticipation of the next Ebola epidemic, we are not ready to deal with it much more effectively than we did last time.

Pharmaceutical companies have put many millions of dollars into this effort, but there has been only one $5 million purchase of one as-yet-unlicensed vaccine for emergencies by Gavi. This underscores why we must have some sort of public subsidy.

We cannot expect profit-making companies to bear this kind of huge risk.

Jeremy Farrar, MD, PhD, the director of the Wellcome Trust, was one of the clear and compelling voices of leadership throughout the Ebola crisis. He says, "As Ebola infection rates have come under control, it's a huge concern that complacency sets in, attention moves to more immediate threats, and Ebola vaccine development is left half-finished."

If that happens, then next time around, the media and congressional committees will demand to know why there is no vaccine when we had plenty of warning from the 2014–15 wake-up call.

Once an effective vaccine or a range of vaccines is proven and licensed, a stockpile should be manufactured. But even more important, certain individuals in likely outbreak areas should be vaccinated ahead of time. These include healthcare workers, ambulance drivers, police officers, and other public safety officials, as well as burial teams. Enough vaccine doses should be prepositioned so that ring vaccination can be carried out immediately once an outbreak is discovered, with sufficient additional doses quickly available to cover an entire afflicted area. I think it is reasonable and rational to secure up to 100 million doses of an effective Ebola vaccine.

I have pushed hard with our CEPI efforts (discussed in chapter 8) to make Ebola vaccine our first "game-changing victory." We can do it; I'm certain of that. We can take Ebola off the table as a major epidemic threat, even if it does mutate into a disease that can be transmitted by just breathing shared air with an Ebola case. The question is, Do we have the collective vision, leadership, and financial support to complete the job?

I believe it was Winston Churchill who said, "It's no use saying, 'We're doing our best.' You have got to succeed in doing what is necessary."

CHAPTER 13

SARS and MERS: Harbingers of Things to Come

An' the dawn comes up like thunder outer China 'crost the Bay!

— Rudyard Kipling, "Mandalay"

In late February 2003 Johnny Chen, a previously healthy forty-seven-year-old American businessman based in Shanghai, was returning to Singapore from Hong Kong when he was stricken with a high temperature and trouble breathing. The flight diverted to Hanoi and he was taken to the French Hospital.

It so happened that Dr. Carlo Urbani, an infectious and tropical disease specialist and president of the Italian chapter of Médecins Sans Frontières, or Doctors Without Borders, was working for the WHO at the Hanoi hospital at that time. He had earned the esteem of his colleagues fighting endemic diseases in Vietnam and Cambodia. And when that venerated organization was awarded the Nobel Peace Prize in 1999, Dr. Urbani was one of those who accepted the award in a ceremony with the king of Norway on December 10 of that year—the anniversary date of Alfred Nobel's death. With some of the prize money, Urbani decided to create a fund to provide critical medicines for the world's poor.

Despite what the other physicians thought—influenza was high on the list of possibilities—Urbani realized that Mr. Chen did not have the typical clinical picture of influenza, since he did not become seriously ill until a week after developing fever and diarrhea.

Urbani treated him with antibiotics and all of the normal support care available at any modern and well-equipped hospital. But nothing worked, and that was when Dr. Urbani grasped that this illness was different from anything he could recall in his career.

After seven days on a ventilator, Johnny Chen was medevaced to Hong Kong. Despite first-rate urgent care, he died on March 13. Back in Hanoi, Urbani realized his worst fear: Other patients in the hospital, and then healthcare workers, were developing the same illness. Chen had infected at least thirty-eight individuals. Urbani contacted the WHO headquarters in Geneva, then sealed off the hospital in an effort to contain whatever the mysterious infectious agent turned out to be.

The story really began a few months earlier, with what looked like an unusually severe flu in Guangdong Province, China—an all-too-frequent starting point for the world's yearly influenza strains. In November 2002, Dr. Klaus Stöhr, manager of the WHO's influenza program, attended a routine meeting in Beijing regarding China's vaccination program. He was told by a Guangdong healthcare official that several people in his region, near Hong Kong, had died from a severe influenza virus. This was the time of year when the influenza sleuths are on high alert for new strains to emerge out of China and the Far East—the world's largest concentration of humans living in close contact with enormous populations of pigs, poultry, and aquatic birds such as ducks and geese. These avian species are the natural reservoir for the virus.

On February 10, 2003, ProMED, the Program for Monitoring Emerging Diseases, posted the following inquiry from Stephen Cunnion, MD:

> Have you heard of an epidemic in Guangzhou? An acquaintance of mine from a teachers' chat room lives there and reports that the hospitals there have been closed and people are dying.

Over the course of the next six months, ProMED's ongoing coverage of this outbreak would play a key role in the world's understanding, identification, and control of another new human pathogen.

Klaus Stöhr took viral samples back to Geneva from his November trip to China, and when the lab analysis showed only a normal flu virus, everyone lowered their guard. By February 2003, though, severe pneumonia cases were showing up in the region around Hong Kong. This time, blood and saliva samples showed no evidence of influenza. "We stopped wondering and started worrying," said Dr. Stöhr.

That was when a number of other experienced public health experts around the world were called in to add our thoughts about what was going on. I remember conference calls that included participants from Hong Kong, Southeast Asia, the WHO in Geneva, the CDC in Atlanta, the NIH in Bethesda, and the HHS incident command center in Washington taking place every day. When I heard the description of the way this unknown disease had suddenly broken onto an unwary public, I thought of the line "An' the dawn comes up like thunder outer China 'crost the Bay!" from the poem by Rudyard Kipling. This outbreak really had come like thunder from China to Hong Kong and Vietnam.

Although there seemed to be a cast of hundreds on the many conference calls being organized by the WHO, I was impressed by how Stöhr and Dr. David Heymann, an American who, at the time, was executive director of the WHO Communicable Diseases Cluster, coordinated all the international investigation activities. In the early days of the outbreak, the "unknown status" of the cause clearly added an elevated sense of concern. Heymann's effort to get multiple labs from around the world working together as a single team was one of the WHO's finest hours.

I remember listening to Carlo Urbani on one of these telephone conferences. While he didn't say much, when he did speak, he didn't sound good. He had become ill while traveling to a medical meeting in Bangkok and was hospitalized upon his arrival. During the first several days of his hospitalization, Urbani would join the WHO international call from his hospital room, where he was in isolation. He had developed a worrisome cough that seemed to be growing steadily worse. Because of the scope of these calls, that cough literally could be heard around the world. In retrospect, I realize it was his most vivid warning that this was something we all needed to take extremely seriously.

On March 29, 2003, he crashed and died, after eighteen days of intensive care in a hospital in Bangkok. He was forty-seven years old. Toward the end, he asked for a priest to administer last rites and directed that samples of his lung tissue be saved for scientific analysis. I fervently hope that Carlo Urbani will be remembered as one of the great heroes of modern epidemiology—a man with a noble mission who sacrificed his own life to care for others and alert the world to a vicious and imminent threat.

In suppressing reporting, China forfeited a critical opportunity to contain the disease in its earliest stage and later apologized to the WHO.

Disease detective work established that the mysterious disease had quietly slipped into Hong Kong on February 21 when sixty-four-year-old Dr. Liu Jianlun arrived from Guangdong to attend a wedding. He had treated patients back home with severe atypical pneumonia. He stayed in Room 911 of the Metropole Hotel, across the hall from Johnny Chen. The next day, he felt sick enough to seek help at the emergency department of Kwong Wah Hospital and was admitted to the intensive care unit. By the time Hong Kong health authorities realized they had a dangerous new infection on their hands, the disease had already begun to spread in Singapore and Vietnam, where Urbani picked it up and sounded the alarm.

By February 25, Dr. Liu's fifty-three-year-old brother-in-law was showing symptoms. He was admitted to Kwong Wah Hospital on March 1. Liu died there on March 4 and his brother-in-law on March 19. That same day, a businessman who had also been in Guangdong flew home from Hong Kong to Taipei, Taiwan, carrying the outbreak to that island. All told, about 80 percent of the Hong Kong cases were traced back to Dr. Liu, including sixteen at the Metropole.

No one yet knew what this terrifying new disease was or where it would strike next. That answer would soon come. On March 5, Sui-chu Kwan, seventy-eight, died at home in Toronto, Canada, of respiratory distress. Like Johnny Chen, she had been a guest at the Metropole during Liu's visit. Two days later, her son Chi Kwai Tse was taken by paramedics to Scarborough Grace Hospital with severe breathing difficulty. He died there six days later.

As reported in the *Globe and Mail* of Toronto, an EMS supervisor named Bruce Englund went to Grace's emergency department the night Chi was brought in after a worried call from the EMS team. Englund contracted his disease there. Fortunately, he survived, but ten years later he was still plagued by chronic fatigue and respiratory problems.

Though no one knew it at the time, bringing Chi to the hospital would trigger the SARS outbreak in the network of Toronto-area hospitals, where it would go through at least six generations of transmission.

On March 12, the WHO issued a global alert, describing an atypical pneumonia, characterized by "severe, acute respiratory syndrome of unknown origin." By March 16, that symptomatological explanation became its name: severe acute respiratory syndrome, or SARS. Two days before that, health authorities in Vancouver, British Columbia, identified a fifty-five-year-old man with the condition who had also been at the Metropole Hotel. He survived, and SARS never broke out on Canada's west coast as it had in Toronto.

By April, the CDC and Canada's National Microbiology Laboratory had identified the SARS virus as a previously unknown coronavirus. Coronaviruses are so named because under electron microscopy the proteins projecting out from the virion's surface resemble a corona. By May, it had been determined that two of the prime reservoirs for the disease were masked palm civets and ferret-badgers, native to the Guangdong region and sold in local markets there as food. So the transmission to humans was probably similar to that of Ebola when locals in rural west-central Africa ate infected bushmeat. Further research indicated that the civets and badgers had most likely caught the virus from bats sometime in the months to years before the outbreak.

The great alarm at this time was that the disease, for which there was no vaccine or specific treatment, would gain a permanent foothold in the human population, as HIV had, or that it would become a seasonal threat like flu.

Fear permeated the region and some nurses chose to resign rather than care for SARS patients, recalling the reaction of some hospital workers to early AIDS victims. The *Toronto Star* ran a front-page headline on March 24 declaring, "Mystery Bug Shuts Hospital Emergency Room." Because so little was known, official communications were often vague or contradictory. The exchange of information between officials and frontline responders was far from systematized and sometimes nonexistent.

On April 2, the WHO issued a recommendation that travelers not go to Guangdong or Hong Kong unless absolutely necessary. On April 23, Toronto was added to that advisory.

What eventually stopped the spread was not high-tech medicine, since there wasn't any specific treatment for SARS. Instead it was implementing impeccable infection control, including isolating patients and making healthcare workers wear protective gear, and then intensive follow-up of both healthcare workers and community contacts, with immediate isolation if they showed any early symptoms of SARS. By mid-May, it looked like the outbreak had tapered off, and Ontario lifted its state of emergency. Within days of the proclamation, the hospitals started filling up again with infected patients. Containment efforts went back into a full-court press, and it took another five weeks before SARS was truly under control in Toronto.

Perhaps the greatest medical mystery of the SARS outbreak was why some people, like Dr. Liu and Mr. Chen, gave the disease to so many people they encountered, even casually, while others who caught it became sick themselves but were hardly infectious at all to others. For reasons we still don't completely understand, certain individuals with coronavirus become "superspreaders."

In the public health–infectious disease world, we worry most about diseases that have high mortality rates and that can be effectively transmitted via the respiratory route—in other words, killer diseases that you can catch just by being in the same air space with an infected person or animal. For most infectious diseases, the likelihood that someone will transmit an infection to another person is called the reproductive rate, or Ro. That rate tends to be fairly similar for cases of the same disease when all of their contacts are vulnerable, that is, have not been vaccinated or previously had the disease. For example, the typical Ro for measles, a highly infectious respiratory infection, is eighteen to twenty. So each case, on average, will transmit the virus to eighteen to twenty susceptible contacts. For poliovirus, which is transmitted by the fecal-oral route, the Ro is usually four to seven.

Superspreaders break the reproductive rate rule. They transmit to many more contacts than other cases with the same infection. It's unclear why superspreaders infect such a large number of those exposed. What we do know is that superspreaders can make coronavirus infections in humans into a very scary situation. These superspreaders are not obvious; they are not necessarily sicker, immunocompromised, older, or pregnant—all the conditions we normally associate with being more infectious.

Altogether, SARS claimed 44 lives in Canada out of a total of 438 probable cases. Globally, the estimated mortality was 916—11 percent of those infected. This is a pretty terrifying mortality rate for an infectious disease with global transmission potential. Toronto's loss of tourism was estimated to have cost around $350 million, with another $380 million loss in retail sales.

The World Bank has estimated that the SARS epidemic caused an estimated $54 billion in worldwide economic loss. Most of this figure was not from direct care costs, but from "aversion behavior" on the part of the public.

As Dr. Anne Schuchat, principal deputy director of the CDC, put it, "The only tools we had to control SARS were ones we've had for hundreds of years." Even so, two very different public health–based activities played critical complementary roles in stopping the SARS outbreak: first, elimination of the animal sources in China, and second, effective infection control. Once the civets and badgers were recognized as the likely source for transmitting the virus to humans, they were removed from the markets in southern Asia and people were warned not to eat or have any contact with them. This was, in a sense, "pulling the pump handle," similar to what Dr. John Snow did with the Broad Street pump in London in 1854.

With no additional infections occurring in humans as a result of animal exposure, all that was left to do was use infection control in hospitals and closely follow contacts of cases in the community to stop them from transmitting to other humans.

If a case contact showed any signs of an early SARS-like illness, he or she was immediately isolated from others. While that was more difficult to do than might be expected, particularly when dealing with superspreaders, human-to-human transmission was stopped and the public health control measures finally succeeded. By the summer of 2003, SARS was extinguished worldwide.

But Dr. Peter Daszak, a disease ecologist and president of EcoHealth Alliance, a global organization dedicated to innovative conservation science linking ecology and the health of humans and wildlife, recently observed, "SARS is alive and well and living in China, and ready for the next outbreak."

Two recent studies support that conclusion. Bats sampled in China and Taiwan were found to be carrying a coronavirus that was genetically almost identical to the SARS virus and that any day could be transmitted to another animal species that has substantial human contact. What happened in Guangdong Province in China in 2002 and 2003 could happen all over again if one of these bat viruses infects humans, most likely via another infected animal. We can't for a moment believe that the SARS virus obituary has been written.

Once we understood the natural history of SARS and coronavirus in wildlife and understood that bats were a likely reservoir, there was no logical reason to suppose that exterminating a bunch of civet cats and ferret-badgers would stop Mother Nature from throwing additional coronaviruses at us.

In the summer of 2012, a man from the Kingdom of Saudi Arabia developed symptoms highly reminiscent of SARS, including severe pneumonia not caused by the usual bacteria and viruses, as well as kidney failure. Two months after the patient's illness, Dr. Ali Mohamed Zaki, an Egyptian microbiologist working in Saudi Arabia, isolated the virus from the man's lung tissue and identified it as a coronavirus similar to SARS. But it wasn't exactly SARS. And like the SARS virus a decade before, this strain was previously unknown. In September, a forty-nine-year-old living in Qatar presented with similar symptoms. His disease turned out to be the same virus. Throughout that fall and winter, additional cases cropped up in Saudi Arabia and Qatar.

The new disease was labeled Middle East respiratory syndrome, or MERS. Retrospective analysis suggested that the first MERS case might have occurred in Jordan in April 2012. The original reservoir for the disease, as far as we can tell, is a bat species found in the Middle East. The bats then transmitted it to dromedary camels—the one-humped species common throughout the Middle East and North Africa. Recent studies have been conducted testing stored blood samples collected from camels in both Africa and the Arabian Peninsula for antibody to the MERS virus or a MERS-like virus. They found that these viruses have been circulating in camels in both areas for at least five years.

It's possible that camels became infected from eating figs and other fruit that the infected bats fed on and that fell to the ground. Contact with bat droppings probably had some role as well. Once the camels became infected, they spread it to other camels and to humans.

The bad news was that it appeared to have a mortality rate even higher than that of SARS—somewhere between 30 and 40 percent—prompting some in the public health community to refer to it as "SARS on steroids." The somewhat better news was that it didn't seem to transmit between people very well. To catch it, you had to have close extended contact with an infected person. Within a few months, though, we would learn that, like SARS, MERS "selected" certain individuals as superspreaders, and it was unpredictable just who those superspreaders would be.

The billion-dollar question is, Where did the MERS virus arise that is now causing deadly human disease? Did it just recently jump to camels, which then transmitted it to humans?

Or was a similar virus endemic in camels for a long time and then somehow took on more dangerous characteristics through mutation? In this latter possibility, many camels could test positive for antibodies to a MERS-like virus, but only those that have become infected with *the* MERS virus are going to pose a risk to humans.

The camels carry the MERS virus but are often asymptomatic. Sometimes they experience a mild respiratory illness. They can become chronically infected, meaning that they may shed the virus for years. However, when they pass it to humans, through respiration or bodily fluids or in their raw milk, humans can develop either a mild illness or life-threatening MERS.

And here is the problem that distinguishes MERS from SARS or any of the other coronaviruses: The virus is now established in the camel population throughout the Middle East; it doesn't even need the bats any longer to propagate.

Now, it's one thing to kill all the ferret-badgers and masked palm civets; no one cares that much about them. Even if you have a real taste for this kind of exotic delicacy, it's not an unreasonable hardship to give it up. But there is absolutely no way you are going to eliminate camels in the Middle East.

Camels are highly prized and virtually sacred in Middle Eastern cultures. Human survival depended on them for thousands of years, and they are still deeply entwined with the way of life and vital to local commerce. Camels are raised for milk, meat, wool, transport, and other work. Their dung is used as fuel. Milk is often the most important camel product and is a staple food of nomads.

Moreover, camels have become an increasingly important agricultural export for the countries in the Horn of Africa. For example, in recent years, Somalia annually has exported camels worth more than $30 million to the Middle East.

Camel racing is a popular sport on the Arabian Peninsula, similar to horse racing in the United States. Winning camels often sell for more than $5 million, and some sell for up to $30 million. And beauty pageants are not just for humans; they are for camels, too. Camel beauty contests, where winning camels fetch prices similar to those paid for racing camels, are gaining popularity.

In short, camel owners are not about to cull an infected herd showing few, if any symptoms, the way the Chinese and Americans have several times culled entire populations of chickens infected with various strains of avian influenza. So we can rule out the idea of eliminating the camels of the Middle East and Africa.

What does this mean for the future of MERS? Well, I fear that MERS is only beginning to show its ugly head. There are more than 1.2 million dromedary camels in the Arabian Peninsula, and 78 percent of them are found in Saudi Arabia, the United Arab Emirates, and Yemen. Bactrian camels, the two-humped variety, reside primarily in China and Mongolia. Africa has an estimated 24 million camels, most of which are in the countries in the Horn of Africa, including Somalia (7 million), Sudan (4.9 million), and Kenya (3.2 million).

If the risk of MERS is related to contact with camels, it makes sense that those countries with the most dromedary camels and most people would have the most human MERS cases. In fact, about 80 percent of MERS cases to date have been documented in Saudi Arabia, a country with only 27.1 million people and 800,000 camels. The human population of the other countries of the Arabian Peninsula is approximately 51 million, with 400,000 camels. The human population in the Horn of Africa region is 225.8 million, with an estimated 16 million camels. Saudi Arabia has only 9.8 percent of the region's human population and 4.3 percent of its camels, but more than 80 percent of the MERS cases. Why? We don't know.

What we do know is that recent studies show that MERS, or MERS-like viruses, have been circulating in camels in the Horn of Africa for some time. Yet there is no evidence of MERS cases thus far in the camel-herding population there. A recently published study found antibodies to the MERS virus in only 2 of 1,122 humans sampled from Kenya. This suggests a relative absence of infection in an African country with a large camel population.

Is it possible that MERS is really a serious public health problem in these countries, and cases are being missed because of their poorly resourced healthcare systems and inadequate disease surveillance? I don't think so. If the MERS virus causing the current outbreaks in Saudi Arabia also occurred in the countries of the Horn of Africa, we would not miss a major superspreader outbreak among other patients and healthcare workers in at least some of the hospitals, even if disease surveillance were poor.

I feel certain that the MERS virus causing serious human disease today emerged in Saudi Arabia or Jordan in the past five to six years. It is likely a mutated strain of the other MERS-like viruses in Africa that do not cause human disease. Since most of the camel trade has been one way—Horn of Africa camels being sold on the Arabian Peninsula—the human disease-causing MERS virus has yet to be seeded in Africa.

But I have little doubt that it will, sometime in the future, as other infectious diseases have, time and time again. It is only a matter of when. Even though most of the general trade goes in the other direction—into Africa—from an epidemiological point of view, it is unreasonable and illogical to suppose that it will not eventually cross the Red Sea.

The next frontier for human MERS will be among the 225.8 million people in the Horn of Africa. In these countries, already lacking many basic healthcare resources, MERS could be devastating. It could be the East African version of what happened with Ebola in West Africa.

I studied the situation in Abu Dhabi at the invitation of the royal family there, giving me the experience of studying MERS at its source in the Middle East as well as where it landed in Korea. I continued to closely monitor the situation in the Middle East and advocate to anyone who had anything to do with vaccines—both camel and human. I told my contacts there that it had become clear to us that the only way to deal with MERS was with a One Health approach that considered both animals and humans. What this means is that while we might come up with a vaccine and/or antivirals that can prevent or minimize the disease in people, the most direct and efficient means of controlling it would be a vaccine for camels and any other mammals found to be carriers. This is the clear strategy to "remove the pump handle" and halt the spread.

MERS has continued to simmer in the Middle East. Between 1950 and 2009, Saudi Arabia had only two ministers of health. Since the emergence of MERS, there have been five, which we strongly believe to be due to their collective inability to control the virus.

At a conference on emerging disease threats at the Institute of Medicine in Washington, DC, in March 2015 (the name changed to National Academy of Medicine on July 1 of that year), I predicted that before long, MERS would show up outside the Middle East—as soon as an unknowing superspreader got on a plane and traveled to a large city. I said I had no idea where or when, but it was almost inevitable.

Less than two months after the conference, a sixty-eight-year-old man returned to South Korea after visiting four Middle Eastern countries. In the nine days between when he started feeling sick and when MERS was finally diagnosed, he had gone to four healthcare facilities. Had his condition been identified early on, he could have been put in isolation and the outbreak might have been stopped in its tracks, or at least better controlled. As it was, by the beginning of June, he had infected more than twenty other people, including family members as well as patients and healthcare workers at two of the hospitals he visited: St. Mary's in Pyeongtaek, and Samsung Medical Center in Seoul.

There is one major reason why the virus spread so quickly in Korea: Inadequate infection control practices were in place, particularly against a highly infectious superspreader. Unfortunately, this same situation is all too common in modern healthcare facilities worldwide.

The economic, social, and political effects were dramatic. Samsung Medical Center closed down to new patients for five weeks—from June 14 until July 20. Nearly 3,000 schools closed. Attendance at sporting events dropped, concerts were postponed, and acts as basic as shopping in stores and supermarkets were curtailed. More than 100,000 trips to Korea were canceled. The Bank of Korea cut its interest rate to a record low and publicly worried that the economy could go into a tailspin. The leadership of President Park Geun-hye became a subject of national debate and she was accused of cutting herself off from the problem.

Health authorities directed all suspected cases to be isolated in hospitals or quarantined at home. Infection controls were reviewed and strengthened. Supermarket shelves were wiped down with disinfectants and subway stations and trains were sprayed regularly. Face masks were recommended in public to avoid respiratory transmission. Altogether, more than 16,000 people were quarantined, including an entire village. The condition of each affected person was officially monitored.

By the end of July, the Korean MERS death toll was 36 out of 186 confirmed cases.

In September, the president of Samsung Medical Center, Dr. Jae-Hoon Song, invited me to come to Seoul with my colleague from the Mayo Clinic, Dr. Pritish Tosh, to evaluate the situation at Samsung and advise him regarding what might be done to avoid a future crisis. I have known Jae for many years and count him as a close friend and highly respected colleague. He is one of the most skilled infectious disease physicians I have ever worked with. Jae had been put in an impossible situation that quickly became both a medical and political crisis. He was hauled up in front of a National Assembly hearing and his emergency department was accused of having missed the original diagnosis of the superspreader and obstructing an epidemiological investigation.

Samsung Medical Center is a major national hospital and on a scientific par with the top regional medical centers in the world. The medical, nursing, and administrative staff number some of the finest and most skilled professionals in the medical field. During the MERS outbreak, many of them performed their jobs in heroic fashion, spending days in the MERS ward, never leaving their desperately ill colleagues and other patients. Contrary to all the rumors, it was at Samsung Medical Center where the index patient was correctly diagnosed after visiting three other facilities. While 285 patients and 193 healthcare workers were exposed to this patient before specific infection-control precautions were implemented, no subsequent transmission occurred at Samsung. The problem initially arose because the index patient exposed thirty-eight people prior to going to Samsung, and one of those, a thirty-five-year-old man who had not traveled outside of Korea, went to Samsung's emergency department, triggering the major spread.

As soon as this individual was considered a possible MERS case, he was placed in isolation, but by that time two additional days had passed and repeated transmission had occurred. Every person with whom he came into any kind of contact while in the ER was examined, interviewed, and tracked.

Today, we are no better prepared to meet a disaster like this than Korea was. If we had had a similar MERS superspreader event in an American hospital, it's very possible we would have had similar results. And I think the public health messages might have been just as mixed as they were during the Ebola outbreak in 2014. Imagine what the media and public reaction would be if the Mayo Clinic, Johns Hopkins Hospital, Massachusetts General Hospital, or the Cleveland Clinic had to be shut down for five weeks or so because of a major MERS superspreader event. It would create a national crisis.

In 2014, a CDC study determined that more than 125,000 people flew directly to the United States from Saudi Arabia and the UAE in a two-month period. Any one of these travelers could be like the sixty-eight-year-old man who returned from the Middle East to South Korea.

In summer 2016, the Samsung Medical Center team responsible for investigating and controlling the MERS outbreak at their institution published a detailed account of their efforts and the lessons they learned in the medical journal *Lancet.* The final paragraph of that article provides a battle-weary conclusion and a voice of experience that the rest of the global healthcare community should take very seriously.

> The potential for similar outbreaks anywhere in the world should be noted from a single traveller as long as MERS-CoV transmission continues in the Middle East. Emergency preparedness and vigilance are critical to prevent further large outbreaks in the future. Our report serves as an international alarm that preparedness in hospitals, laboratories, and governmental agencies is the key not only for MERS-CoV infections but also for other new emerging infectious diseases.

There is no question in my mind that the Korean outbreak will not remain an isolated incident in the natural history of MERS. Wherever it strikes next, hospitals and public health officials will face the same challenges.

So when it comes to MERS, we are facing two big issues. There is no reason to assume that the next outbreak will be confined to one city or region as it was in Korea. If the virus does find its way onto the African continent, it will be extremely difficult to eliminate or even control. We have an opportunity now to do something decisive before that happens, but the window won't stay open indefinitely.

As we were completing this book, the WHO put out a comprehensive document: *A Roadmap for Research and Product Development Against Middle East Respiratory Syndrome: Coronavirus (MERS CoV)*. It defines the critical product development needed to hit MERS head-on. Both human and camel vaccines are top priorities. The road map also prioritizes effective treatments and better diagnostic testing.

MERS vaccine research and development has also been identified as a priority by the Foundation for Vaccine Research, the Norwegian Institute of Public Health, and CEPI. Will a vaccine ever become a reality? I don't know, as there is no prospective pot of money at the end of the vaccine research and development rainbow. Nor is there a Manhattan Project–like authority to direct such an effort. I fear the WHO road map document will grow dusty on office shelves. I have personally experienced this. At CIDRAP, we produced a comprehensive report on the need for game-changing influenza vaccines that has been ignored for years. You will read about it in our final chapter.

The SARS outbreak has left the world one legacy that continues to haunt us today. A number of vaccine research, development, and manufacturing companies stepped forward in the early days of the SARS outbreak in 2003 at the request of the WHO and invested many millions of dollars in SARS vaccine work. I'm not aware that anyone knows exactly how much was invested across the pharmaceutical industry, but it's likely in the hundreds of millions of dollars. The industry wanted to do the right thing by assisting the world in responding to this public health crisis, and to capitalize on an investment opportunity.

When the outbreak was extinguished by the end of the summer of 2003, interest from government agencies and philanthropic organizations to support additional research on a SARS vaccine virtually disappeared. There was no interest registered at the time for ever buying such a vaccine. The companies were left largely holding the bag for the early SARS vaccine research costs. As we've noted, this corporate "memory" remains a major concern going forward for vaccine-related investment.

As of this writing, in the wake of the West African Ebola epidemic, government interest in the disease has waned and vaccine manufacturers have made nothing for their efforts. Given their wariness at being "left at the altar" yet again, we should not expect major players in vaccine manufacturing to cough up big money for the next international infectious disease crisis.

That is our first challenge. If we don't face up to it, and don't heed the recommendations and strategies in these expert reports, I have no doubt that we will regret our inaction.

CHAPTER 14

Mosquitoes: Public Health Enemy Number One

If you think you are too small to make a difference, try sleeping with a mosquito.

— Dalai Lama

I have been involved in one way or another at some point in my career with all of the major diseases we've discussed. As an infectious disease epidemiologist, I can relate to them and to the means by which they transmit. But with mosquitoes and the diseases they carry, it's very personal.

In 1997, we built a house in the western Twin Cities suburbs fronting on beautiful Lake Minnetonka. It was a heavily wooded lot with twenty-nine large red oak trees. My sixteen-year-old son Ryan had been spending the first summer month with his grandparents, north of Minneapolis, and came home to help me plant trees around the new house. At one point, he was digging holes for the planting near the perimeter of the property, and I was watering the newly sodded lawn.

About a week later, Ryan developed a severe headache that he couldn't get rid of. I remember it was a Saturday evening and he and I were watching the Minnesota Twins game, when he said he was just too tired and was going to sleep in his bedroom in the walkout basement.

The next morning I yelled down to him to get up and get himself ready for church. He mumbled something back about still being tired, and that he was going to stay in bed.

When I came back from the service, I called down to him that I was home, but he didn't respond. I went down to his room and found him moaning in a totally incoherent way. There was evidence of projectile vomiting around the room and no indication that he had made an effort even to get to the bathroom.

I had led a large bacterial meningitis outbreak response the previous year among high school students in Mankato, southwest of Minneapolis, and that was the first thing I thought of. A sixteen-year-old boy with classic symptoms similar to Ryan's had died in that outbreak.

No one else was home at the time. I picked him up, put him over my shoulder, and carried him to the front seat of the car. I called ahead to Minneapolis Children's Hospital and then drove there as fast as I could. While driving, I got hold of Kris Moore, my coleader in the Mankato response, and her husband, and they arrived shortly after Ryan and me in the ER.

A lumbar puncture revealed no visible evidence of bacteria, which partially relieved my concerns over bacterial meningitis, but then we started wondering what Ryan might have. He was admitted to the hospital and the next day his condition remained the same. We finally started seeing some improvement late Monday afternoon, and by nighttime, he seemed to be recovering from whatever it was.

Then, Tuesday night he crashed. He was transferred to the ICU and I had to confront the possibility that we could lose him.

Ryan's doctors and I were going through everything we could think of, and given my professional world, I suggested running an antibody test for the mosquito-associated viruses found in Minnesota. Despite my previous experience with it, I really didn't think he had anything like La Crosse encephalitis, because the incubation period was typically longer than a week and there was none of the virus in the area where he'd been with his grandparents prior to that time (or so I thought).

I was surprised when the test came back positive. It made us all rethink the conventional wisdom about the incubation period of La Crosse encephalitis and accept the fact that there were a lot of variables about viruses that we just hadn't faced. I was actually encouraged by the diagnosis, though, because despite the tragic outcome for the first young patient back in 1960, and the reality that we still had no specific treatment, the prognosis for the disease was statistically a good deal better than for some of the others on our differential diagnosis.

Little by little, with the aggressive support therapy of the hospital, Ryan did get better and pulled through without any obvious deficits. I was still worried about residual brain damage, but we'd just have to wait and see.

When the Metropolitan Mosquito Control District staff surveyed the area near my house, they found the tree holes—natural indentations or rotted out sections in the branch crotches of mature trees—I had unknowingly been watering on the edge of my yard every time I watered the lawn. They also found *Aedes triseriatus,* and tests of those mosquitoes showed the La Crosse virus. The tree holes in the entire neighborhood were filled.

The news media picked up this story and it became a cautionary tale of how a high-level public health officer had stirred up mosquitoes by watering his trees and then missed the implications of his actions, even though he'd studied the disease in depth.

Fortunately, Ryan had no residual effects from his encounter with La Crosse encephalitis. Years later, when his sister Erin was in medical school at the University of Minnesota, she had a rotation in neurology. During a presentation on La Crosse encephalitis, she realized that the unnamed patient being described was Ryan.

Americans tend to think of mosquitoes as an annoyance rather than a deadly foe. We protect ourselves with bug spray if we remember, but mostly we're content just to slap them dead while they're midbite. Of course, not all of them are dangerous. There are around 3,000 species of mosquitoes, and only a relatively few are known to transmit disease to humans. But those that do are Public Enemy Number One of the animal world. And as it turned out, a tiny, whining mosquito was the reason my son's life was in danger.

Mosquitoes are arthropods, meaning they have exoskeletons, segmented bodies, and jointed appendages. Different species exhibit different behaviors, important factors in our understanding of vector-borne diseases and how they spread. Some mosquito species can travel many miles a day on the wind. Others will not venture across a rural road. Some live only in wooded areas, others in marshlands. There are those that have adapted to living among us, like rats and cockroaches. They take up residence in our backyards and even in our closets. Some lay their eggs mainly in stagnant drainage areas or tree holes that accumulate water after a rainfall. Others can multiply in a plastic soda bottle capful of water. Any kind of mosquito control program has to be based on knowledge of which species is carrying the virus or parasite.

Unlike the human world, where the vast majority of criminals are male, with mosquitoes, it's only the females that bite, through the slender, hollow, tubelike extension from the mouth, called a proboscis. In some species, the female needs the nutrients in blood to produce eggs, and in others the blood stimulates production of more eggs. When she bites, the mosquito injects saliva into the tiny wound; the saliva contains an anticoagulant that keeps the blood from clogging up her proboscis. The itchy red bump on the skin left after the bite is the result of a histamine compound fighting off the invading protein. It's the saliva that contains the virus or parasite that then infects us. And we're not the only species affected by mosquitoes. Various types will seek blood from humans down to small rodents and even reptiles.

For a mosquito to transmit an infectious agent, it has to become infected. Fortunately, only a small percentage of mosquito species are susceptible to infection with human disease pathogens. The main way they get infected is by feeding on a human or animal that is also infected. For example, early in the summer, a mosquito carrying West Nile or, say, Eastern or Western equine encephalitis virus, bites nestling birds that can't yet fly. These young birds become infected and are now carriers. Other mosquitoes bite these immobile birds and then bite other birds and humans in a pyramid of infection that keeps building on itself.

Malaria, on the other hand, is largely a human disease that is transmitted to biting mosquitoes, which then transmit it back to other humans. More recently we have seen an increase in strains of a malaria parasite that primarily infects monkeys also infecting humans in Southeast Asia.

Temperature plays an important role, too, because it affects the extrinsic incubation period: how quickly the mosquito becomes infected with whatever it has taken in through a blood meal, and then how quickly it becomes infectious. The warmer the temperature, the shorter the extrinsic incubation period for most vector-borne diseases. This is why climate change is such a significant factor in our consideration of transmission.

As it turned out, the particular breed of mosquito that was important in Ryan's case, *Ae. triseriatus,* was one I had a long history with. *Ae. triseriatus* and I go way back.

When I was a sophomore in high school, a local game warden I had befriended helped me land a summer job with the Iowa State Hygienic Laboratory—the state's official public health lab. This was during a time when an increasing number of cases of La Crosse encephalitis were occurring each summer near my hometown of Waukon. This is a nasty virus that causes swelling of the brain, which in turn can cause fatigue, fever, headache, nausea, and vomiting and can lead to seizures, coma, and sometimes paralysis. The severe symptoms occur most often in people under about sixteen years of age, and though they are usually transitory, occasionally they can be permanent or even fatal.

It was originally known as California encephalitis but got its present name when a young girl from Minnesota was treated for an unknown disease at the La Crosse Gundersen Clinic in Wisconsin, which is about sixty miles northeast of Waukon. Tragically, she died. Samples of her brain and spinal tissue were saved, and five years later an arbovirus was isolated from those samples.

La Crosse encephalitis is carried and transmitted by the *Ae. triseriatus:* a so-called tree-hole mosquito that spawns in hardwood trees, containers of water, abandoned tires, and other junk that holds a little bit of rainfall and is out of direct sunlight.

A tree hole occurs in hardwoods like oaks when the trunk and a large branch form a crotch that can hold water from rainfall or watering. The crotch becomes an ideal breeding ground for *Ae. triseriatus.* It's dark and calm, protected from wind, and often a receptacle for leaf litter, which serves as a food source for the microorganisms the larvae feed on.

Ae. triseriatus seldom travels more than a few hundred yards from where it is hatched. The primary reservoir for the disease is rodents, but once a mosquito is infected, transovarian transmission is possible with this particular disease. That is, a newly hatched *Ae. triseriatus* from an infected mother is infected and can transmit La Crosse without having taken an infected blood meal itself.

When I began working on La Crosse encephalitis there were usually between twenty and forty cases of it in northeast Iowa, southeast Minnesota, and southwest Wisconsin each year. Most of them were kids. The first symptoms were usually a headache and stiff neck.

I had a basic lab set up in the basement of my house with equipment the hygienic laboratory had provided me. I had a rudimentary microscope for sorting the insects I collected, and I learned how to identify the thirty or so mosquito species that were indigenous to our area. I had glass vials for the samples and a special dry-ice freezer in which they could be preserved. I also had a number of light traps for catching the mosquitoes each night. The traps consisted of a large netlike bag hanging down from a clear plastic cylinder containing a light and a fan. Each night in the hours before sunset I would go on my ten- to twenty-mile route setting up ten to fifteen light traps. They would run all night with power from a motorcycle battery. I would also hang a cloth bag of dry ice above the light trap; the melting ice gave off carbon dioxide and attracted mosquitoes to the light. Once near the light, the mosquitoes were sucked into the net bag by the fan. Each day just before sunrise I would reverse my route and pick up the bags full of insects. After an hour in the dry-ice freezer the insects were all dead and waiting to be sorted into vials.

My job was to trap *Ae. triseriatus* in woodland areas near where cases of La Crosse had occurred. I'd generally find them in shady locations near their hatching places in tree holes and the crotches of trees where the branches connected, or in the many discarded tires and other nonbiodegradable containers commonly found on Iowa farms. I'd ship the samples off to the state lab every week. In return I'd receive a shipment of dry ice to replenish the freezer and provide the carbon dioxide bait that I needed each night.

As part of my project responsibilities, I also kept rabbits in cages in the same areas as the light traps. Once a week I'd draw blood from them to see if they'd been infected. I had a centrifuge to spin off the serum component of the blood because that is where the antibodies are found. With this assignment and all the lab equipment, I felt like a real scientist.

I was still doing this my junior year in high school and loving it. One Saturday night, I came home late and found my mom crying in the kitchen. I asked her what had happened and she told me my father had come home drunk, as he often did, and he'd gone down to the basement in a rage and smashed up part of my lab. Then he left again. He often slept off his drunken stupors on the floor of his darkroom at the local newspaper office.

The basement was a mess: shattered glass vials all over the place. Fortunately, the dry-ice freezer with the samples was locked so my younger brothers and sisters couldn't put their heads into the freezer and get stuck. The glass lens on the microscope was in pieces. I was angry, stunned, and frightened by what this would mean to my further employment with the state lab. So when my dad came home the next day I confronted him and demanded to know why he had done that, knowing how important the lab and the work were to me.

"What the hell were those goddamned things doing there, anyway?" he shot back. I never did figure out why he had destroyed what he did—maybe it was some deep-seated resentment against me, or disappointment in his own life, that he couldn't bring himself to articulate. This all took place a little more than a year before I threw him out for good.

Monday morning, I had to call Dr. William Hausler, the director of the state lab and a nationally prominent microbiologist. I was terrified that I would lose my job and have to pay for all of the ruined equipment.

I got up my courage to make the call and decided my only approach was to tell him exactly what had happened. This was back in a time and part of the country where things like this were hushed up.

The first thing Dr. Hausler said after I tearfully told my story was, "Are you okay?" I told him I was. "Is your family okay?" he pressed on. Yeah, under the circumstances, they were okay, I replied.

"Equipment can always be replaced," he said. "We'll just deal with whatever happens. Do you think your dad will do it again?"

I said, "I don't know, but I hope not."

My sudden relief, respect, and love for Dr. Hausler knew no bounds. I kept my job, he had the lab replace the broken equipment, and he and I kept in close contact throughout my professional career, until he died in 2011. I had the good fortune to give talks throughout my career where Bill was in the audience. In some cases, he even introduced me. I never missed an opportunity to tell the world my story of Bill and that early career crisis. It was the least I could do to honor the man who gave me my start in this business. And he taught me a lifelong lesson about how to prioritize and act upon the most important values in the workplace. Even though Bill is gone, I will forever be a mentee of his. By the way, my father never touched the lab again.

Mosquitoes continued to be a major concern during my early years heading up the Acute Disease Epidemiology Section at the Minnesota Department of Health. I was closely involved with the follow-up of La Crosse cases in Minnesota, trying to identify and eliminate the breeding sites of the *Ae. triseriatus* responsible for the new cases.

In the early 1980s, we saw major Western equine encephalitis virus activity in birds and *Culex tarsalis* mosquitoes and worked closely with the CDC to prevent major summer outbreaks. *Cx. tarsalis* is one of those mosquito species that breeds in small bodies of water, including marshes and prairie pothole ponds. It can be carried by prevailing winds more than twenty miles in a night.

In 1983, lab tests confirmed Western equine encephalitis virus in an increasing number of mosquito samples as well as cases in horses in the west-central part of the state. On top of this, because of the very warm and wet summer, mosquito populations were at an all-time high. We had all the ingredients for a possible human outbreak. I found myself in charge of an extensive pesticide-spraying program to prevent the disease from taking hold in horses and humans.

We began spraying in thirteen of eighteen targeted communities, employing twelve airplanes, including the crack US Air Force spray team from Wright-Patterson Air Force Base in Dayton, Ohio. Then all of a sudden the Minnesota attorney general's office got word that a judge in Otter Tail County had issued a temporary restraining order at the request of the Minnesota Honey Producers Association and two beekeepers who were concerned about possible harm to their hives. I said we would cover the hives and assume responsibility for any damage. They suggested we spray only from sunset to sunrise, when the bees weren't active.

At midnight that same night, the chief justice of the Minnesota Supreme Court convened the entire court in a conference room at the state health department. I had not slept in the previous forty hours, yet I found myself the only witness representing the state of Minnesota. After hearing testimony from me and from representatives of the other side, the court lifted the restraining order. We agreed not to spray between ten a.m. and five p.m. and to stay as close to our targeted areas as possible. This was a classic case of weighing the best public health interests against the legitimate concerns of private citizens and businesses, and I think we tried to work it out with consideration for all sides.

What resulted was one of the largest aerial spraying efforts to control Western equine encephalitis ever undertaken in the United States. It covered forty counties, or roughly half the state population, cost $1.7 million, involved a broken hose from one contract airplane that dumped 400 gallons of the chemical on a farm barnyard, and prompted about a hundred lawsuits for damage, for which the health department paid out a total of around $59,000.

But there was no outbreak, and when a reporter questioned

me, I said that under the same circumstances, I'd do it again. We were never sure if a human outbreak would have occurred if we hadn't sprayed. That's the challenge of proactive public health practice. If you prevent something from happening because of your actions, you'll always be second-guessed as to whether the action was necessary. On the other hand, if you don't act on the information you have and an outbreak occurs, you will be burned at the stake by the media, elected officials, and even your colleagues. I have always taken the position as a public health professional that I'd rather have to answer for something I did than for something I didn't do.

Ultimately, the honey producers supported us despite the loss of some hives, and the CDC issued a statement that read, "The program for the containment of the Western equine encephalitis threat in Minnesota was excellent."

Two years later, the CDC asked me to be part of a working group on *Aedes albopictus,* the mosquito that transmits dengue and yellow fever. It was chaired by William "Bill" Reeves of Berkeley, one of the giants in the field of vector-borne diseases, who had consulted with us on our spraying program in Minnesota and was one of the main reasons I had confidence the program would work.

This was one of those all-too-unusual situations in which we tried to be proactive rather than reactive. Though they weren't yet spreading any vector-borne disease, *Ae. albopictus* had been identified in the United States for the first time and the CDC wanted to get ahead of the problem. It turned out that large numbers of retreaded truck tires were being imported from the Far East. Before they were loaded onto ships, many of those tires were lying around before and after retreading—perfect collecting vessels for rainwater, making them perfect receptacles for mosquitoes laying their eggs. This is the way so many infectious diseases spread. *Ae. aegypti,* the "cockroach" of mosquitoes due to its ability to live quite nicely in the human environment, inside or outside, first came to the Americas by hitching rides from Africa on slave ships. The pursuit of public health almost always involves the study of unintended consequences.

Ae. triseriatus remains an important public health challenge. But *Ae. aegypti* is the cause of a current global public health crisis.

As early as 1915, the Rockefeller Foundation made the research and eradication of yellow fever a priority. That put *Ae. aegypti* in the bull's-eye of public health, as it is the primary vector of yellow fever. In the late 1940s, Fred Soper of the Rockefeller Foundation and the Pan American Sanitary Organization (later to become the Pan American Health Organization) launched a coordinated and comprehensive effort to eradicate *Ae. aegypti* throughout the Americas. The program developed strong national efforts that used a combination of elimination methods, including breeding-site reduction and killing both the mosquito larvae and the adults through the use of pesticides like DDT (dichlorodiphenyltrichloroethane).

In a real sense, we succeeded so well that we considered the problem solved and began to take the elimination of mosquitoes for granted, leading to apathy and a lapse of vigilance. The advance of nonbiodegradable products that often ended up littering our outdoor environments didn't help, either.

Throughout the 1960s and 1970s, the proliferation of megacity slums around the developing world meant a proliferation of casually discarded plastics and solid waste—perfect breeding grounds for *Ae. aegypti*.

Now, not only have we lost the ground we gained; we've gone backward. For some of the mosquito-borne diseases, such as those for which *Ae. aegypti* is the primary vector, the rate of human infection is higher today than it was at any time in human history. This is surely true for the current grand slam of yellow fever, dengue, chikungunya, and Zika.

The truth is that no country today is adequately controlling mosquitoes, particularly the *Aedes* species. But in the not so distant past, we did achieve major control of *Ae. aegypti* in the Americas. This effort began shortly after the turn of the twentieth century, with an emphasis on source reduction—finding where the mosquitoes lay their eggs and eliminating those sites. By 1962, a sizable portion of the Western Hemisphere had been declared totally free of the mosquito and of dengue. That was about the time our path to failure began. To better understand that failure, we have to understand the successes of the past.

In the Marianao district of Havana, Cuba, there is a tall stone monument with a sculpted syringe on top in memory of Dr. Carlos Finlay.

The National Military Medical Center in Bethesda, Maryland, is named for Dr. Walter Reed.

The Gorgas Medal, named for Dr. William C. Gorgas, is bestowed by the Association of Military Surgeons of the United States.

These and numerous other much-deserved honors testify to the greatness of these three pioneering giants of infectious disease and the ongoing war against the *Ae. aegypti* mosquito.

If it weren't for *Ae. aegypti,* the French might have succeeded in building the Panama Canal, instead of abandoning the project after a thirteen-year effort that saw as many as 200 workers a month die of yellow fever and other vector-borne diseases. And it was the leadership in sanitation and mosquito control of Gorgas, building on the theories and discoveries of Finlay and Reed, which allowed Americans to finish the job and revolutionize shipping and commerce in the Western Hemisphere.

Yellow Fever

Yellow fever—so named because in its severe form it damages the liver and causes jaundice—is a flavivirus believed to have originated in East or Central Africa. Most people who are infected have mild symptoms or none at all. The most frequently reported effects include the sudden onset of fever, chills, severe headache, back pain, general body aches, nausea and vomiting, fatigue, and weakness. Most people improve after the initial presentation. But after a brief remission of hours to a day, roughly 15 percent of cases progress to develop a more severe form of the disease, characterized by high fever, jaundice, bleeding, and eventually shock and failure of multiple organs. There is no specific treatment for severe yellow fever. Twenty to 50 percent of severe cases will die.

Its prime vector, *Ae. aegypti,* came to the New World on slave ships and caused the first recorded outbreak in 1647 on the island of Barbados. It gradually traveled up and down the Caribbean and East Coast until it reached New York in the 1660s and Recife, Brazil, in 1685. A major outbreak hit Philadelphia and the Mississippi River valley in 1669. It wasn't long before no warm-weather region of the Americas was immune from *Aedes*'s relentless colonization.

Carlos Finlay was a Cuban physician educated at Jefferson Medical College in Philadelphia, where he became acquainted with Dr. John Kearsley Mitchell, a major proponent of germ theory, the intellectual foundation of infectious disease medicine. Finlay returned to Havana in 1857 and established a practice in ophthalmology. But it is not eyes for which Finlay is remembered. It was his theory that the scourge of yellow fever was caused not by the "bad air" of the miasma theory or even by human-to-human contact, but through the bites of the prevalent mosquito population. He presented his theory at the 1881 International Sanitary Conference in Washington, DC. A year later, he upped the ante by identifying the *Aedes* genus as the prime culprit and suggested that control of mosquitoes could go a long way toward stamping out yellow fever and malaria.

In June 1900, Major Walter Reed, MD, of the US Army Medical Corps was assigned by the army's surgeon general, George Miller

Sternberg, to go to Cuba in the aftermath of the Spanish-American War to test out Finlay's concepts. Reed had, for the time, a strong background in infectious disease research, having extensive experience with typhoid fever outbreaks in military outposts.

On the outskirts of Havana, Reed had constructed two barracks-like buildings, which he dubbed Fomite House (fomite being a physical object that can carry and, when touched, transmit infection) and Mosquito House. Volunteers were offered money to sleep in one of the two buildings. Fomite House was truly disgusting, with dirty bed linens contaminated with the vomit, urine, and feces of previous yellow fever sufferers. Accounts recall visitors puking just upon entering the fetid atmosphere. But Reed made sure no mosquitoes got in.

Mosquito House, by contrast, was kept spotlessly clean, with good air circulation. Inside, a sleeping space was divided by a partition that went from the floor to the roof. One side was kept completely free of mosquitoes. On the other side, the bugs were intentionally introduced.

At the end of the experiment, none of the volunteers in the mosquito-free section of the clean house or the sorry souls who had occupied Fomite House had come down with any serious illness. But many of the volunteers in the mosquito-infested section came down with yellow fever.

This was the proof the army and the rest of the medical community needed. General Leonard Wood, a well-respected physician in his own right and, at the time, military governor of Cuba, proclaimed, "The confirmation of Dr. Finlay's doctrine is the greatest step forward made in medical science since Jenner's discovery of the [smallpox] vaccination."

Reed's work, for which he freely credited Finlay, led to mosquito control in the tropics and a tremendous decline in the mortality rates for yellow fever. This, in turn, led to Gorgas's success in controlling yellow fever in Florida, Cuba, and then Panama.

Roughly from that time on, mosquito control became a

national priority, with the federal government taking on a leadership position. Through the 1940s and 1950s, an international effort spearheaded by the Pan American Health Organization and the Rockefeller Foundation essentially eliminated the *Ae. aegypti* mosquito from twenty-three Western Hemisphere countries.

By the 1960s, *Ae. aegypti* was almost eliminated from the Americas due in part to the extensive DDT spraying in homes. The formula won a 1948 Nobel Prize for Paul Hermann Müller, its Swiss chemist inventor. But after Rachel Carson's 1962 book, *Silent Spring,* raised environmental awareness and called into question the environmental and physiological effects of DDT, the insecticide was gradually banned and withdrawn.

Since its publication, *Silent Spring* has been an endless source of discussion and controversy, and it is not our aim to argue its accuracy or even its legacy, one way or the other. It should be noted, however, that the extensive agricultural use of DDT, rather than the extremely limited public health use, is what drove the environmental effects and the resulting movement against it. But by 1970, several years into the *Silent Spring*/DDT-banning era, the public health community declared victory over *Ae. aegypti* and moved on to other priorities.

Suffice it to say that in the years since the end of DDT spraying, *Ae. aegypti* and other mosquito species have crept—actually buzzed—back into human environments and used the three-decade era of complacency at the end of the twentieth century as an opportunity to regroup and once again flourish. Today, *Aedes* is largely resistant to DDT, making its use moot.

Dr. Duane J. Gubler, emeritus professor at the Duke–National University of Singapore Medical School, is one of the world's leading experts on vector-borne diseases. He has identified four drivers that, combined with the post-1970 apathy regarding *Aedes,* have led to the worldwide problem we face today: unplanned urbanization and population growth; globalization, with modern air transport and increased international travel; the modern solid waste challenge (nonbiodegradable garbage made of plastic and rubber, which become ideal *Aedes* breeding sites); and lack of effective on-the-ground mosquito control. Together, these factors have allowed *Ae. aegypti* to adapt to living in crowded human populations. It has moved around the world with ease through modern passenger transportation and shipping, and it thrives in any environment where humans do.

Yellow fever, whose conquest defined one of public health's great triumphs, is now back. For now it remains largely a disease of the African continent, with an estimated 180,000 cases of serious illness, including fever and jaundice, every year. Of these, an estimated 78,000 will die. But according to Gubler, it's only a matter of time before it becomes reestablished in the Western Hemisphere tropics and warm regions.

In a 2011 medical journal editorial, Gubler said he expected to see yellow fever cases crop up in megacities throughout the developing world. If that happened, he wrote, "the virus would move very quickly...causing a global health emergency." He went so far as to warn, "The world is sitting on a 'time bomb' with yellow fever, which is a more virulent virus than dengue."

That time bomb may have arrived. In December 2015 Angola notified the WHO of an emerging outbreak of yellow fever, exactly what Gubler had been worrying about. There has been extensive local transmission in Luanda, the capital city, with a population of more than 7 million. The epidemic has spread to several other major urban areas in the country.

Yellow fever has broken out across a broad band from Senegal all the way south to Angola on the west coast, and across the continent to Sudan, South Sudan, Uganda, Ethiopia, and Kenya. The WHO declared a Grade 2 (of 3) emergency in March 2016. While the disease seemed to have been under control in Angola and the Democratic Republic of the Congo by summer 2016, only time will tell if the control will actually end this crisis.

The public health experience in Angola highlights the management challenges. The WHO shipped more than 6 million doses of yellow fever vaccine a month before declaring the emergency. By the end of March, about 1 million of those doses had inexplicably disappeared. Some of the remaining doses were sent to areas unaffected by the disease, and large quantities were shipped without syringes, making them unusable. An Associated Press report stated, "This lack of oversight and mismanagement has undermined control of the outbreak in Central Africa, the worst yellow fever epidemic in decades."

The outbreak in the Democratic Republic of the Congo, focused in Kinshasa, could turn into an explosive megacity epidemic. If this happens, the likelihood of spread to Asia and the Americas is substantially increased. Imagine a yellow fever epidemic in the Americas right on the heels of chikungunya and Zika and overlaid on dengue.

Expansion of yellow fever to China has become chillingly plausible. Dr. Sean Wasserman of the University of Cape Town, South Africa, was lead author of an article entitled "Yellow Fever Cases in Asia: Primed for an Epidemic," published on May 5, 2016, in the *International Journal of Infectious Diseases.* He and his two coauthors warned:

> The current scenario of a yellow fever outbreak in Angola, where there is a large Chinese workforce, most of whom are unvaccinated, coupled with high volumes of air travel to an environment conducive to transmission in Asia, is unprecedented in history. These conditions raise the alarming possibility of a yellow fever epidemic, with a case fatality of up to 50%, in a region with a susceptible population of two billion people and where there is extremely limited infrastructure to respond effectively.

Apart from the newly licensed dengue vaccine, of all the *Aedes*-transmitted diseases, yellow fever is the only one for which there is an established, effective, and inexpensive vaccine. But there is a real problem. We don't, and won't, have enough vaccine to cover even a small percentage of those who will need it, should large cities in Africa require immediate vaccination due to an expanding outbreak. And if urban yellow fever cases show up in either the Americas or Asia, the situation only becomes more serious.

How did this happen? Why aren't we more prepared?

The yellow fever vaccine is highly effective; a single dose provides lifelong protection to most recipients. But it is what we call a "legacy" vaccine, meaning it's old by modern vaccine standards and one of the more difficult ones to make. Like most of our influenza vaccine stock, it is made in embryonated chicken eggs, with production methods that haven't changed substantially in the past eighty years. It takes up to six months to produce the vaccine and it is vulnerable to manufacturing problems.

There are only six manufacturers of yellow fever vaccine, and they can turn out only 50 to 100 million doses per year. Two of the manufacturers produce only enough vaccine for their own in-country use. Remember that there are more than 3.9 billion people living in an area of the world with thriving *Ae. aegypti* populations. It is simply not possible to suddenly gear up production facilities and make more vaccine quickly, even if money were not an issue. It's like building a skyscraper; no matter how much you're willing to put into the process, you can put on only one story at a time.

It would take years to bring on more production capacity. Unfortunately, things are going to get worse with our current production capacity at a most inopportune time. One of the six major manufacturing facilities was shut down in 2016 for renovation.

Despite warnings over the years about the future of *Aedes*-related diseases throughout the world from people including Gubler, me, and others, we are not even close to being ready to take on a rapidly emerging global yellow fever outbreak with our current vaccine. But there is one possible sliver of hope. Studies have shown that the current vaccine could be diluted to one-fifth or even one-tenth of the current dose and still provide good protection. A number of yellow fever experts agree. The WHO approved this approach in June 2016, but it's not a slam dunk. There are still concerns about whether a diluted vaccine will be stable and work equally well in children and adults. And, even with maximum vaccine dilution, we wouldn't have enough to cover the populations at risk for an emerging epidemic of yellow fever in Africa, Asia, and the Americas. Yellow fever is the one vector-borne disease that could take off globally and make Ebola and Zika morbidity and mortality take a backseat. We now live in an *Aedes* world. Even if this current African outbreak doesn't ignite a global urban-based epidemic, we can be sure that another one will.

Dengue

Dengue is currently the most important vector-borne virus disease that affects humans. It comes in two forms: Dengue fever is a flu-like illness, largely without complications and with a predictable recovery. Dengue hemorrhagic fever (DHF), on the other hand, is a relatively new disease and can lead to death. While there is some debate in the scientific circles as to the magnitude of the problem, a 2013 study from a number of leading academic institutions, including the University of Oxford, Harvard University, and the University of Singapore, concluded that there are approximately 390 million dengue infections annually, most of which have no or very mild symptoms. But there are at least 96 million that have more severe symptoms. In Southeast Asia, DHF is one of the leading causes of hospitalization and death of children.

"Dengue" is a Spanish word of unknown origin, but it may have derived from the Swahili phrase *kidinga popo,* denoting a disease caused by an evil spirit. Founding Father Dr. Benjamin Rush called it both "breakbone fever" and "bilious remitting fever." Many patients present with symptoms such as fever, rash, and muscle and joint pain—sometimes making it feel as if your bones are breaking.

The four "serotypes," or distinct versions, of the virus are classified as DEN-1 through DEN-4. Major epidemics of dengue—particularly DHF—in large tropical urban centers, caused by all four serotypes, result in significant morbidity and mortality, especially in resource-poor countries, where they often cause a breakdown in primary healthcare and create chaos as hospitals and clinics become overloaded with patients.

While exposure to any one of the serotypes likely imparts permanent immunity to that particular one, it is not cross-protective with any of the others. DHF can occur if an individual is then exposed to another of the serotypes. DHF is characterized by severe internal bleeding, sudden drop in blood pressure leading to shock, and, far too often, death. It's a condition known as immune enhancement disease. Having some antibody from another dengue strain causes the patient's own immune system to overreact and results in this life-threatening disease. Love may be lovelier the second time around according to the 1960s' song, but dengue definitely is not.

This is a relatively new development in the natural history of the disease. Dengue has been known for nearly 1,000 years, first identified during the Jin dynasty in China, where it was already associated with flying insects. In 1907, it was the second infectious disease, after yellow fever, confirmed to be caused by a virus. But it was during World War II that dengue evolved into the threat we know today.

Thanks to the mass transportation of troops across Asia and

the Pacific, the resultant disruption of the local ecology, and then the rapid postwar urbanization of Southeast Asia, the different serotypes spread, and cases of more severe forms of the disease emerged, first reported in the Philippines and Thailand in 1953. By the 1970s, it had become a significant cause of child mortality throughout the Pacific region. What we now call dengue hemorrhagic fever was seen, beginning in Central and South America in the early 1980s, in the form of DEN-2, in patients who already had antibodies for DEN-1.

The WHO has set a goal of reducing dengue morbidity by at least 25 percent and mortality by at least 50 percent by 2020. Whether we can meet this goal depends largely on the development of effective vaccines. The first, CYD-TDV, was initially licensed in Mexico in December 2015 by Sanofi Pasteur, the vaccine division of Sanofi pharmaceutical company. Phase III clinical trials showed an average efficacy between 40 and 50 percent for DEN-1, 30 to 40 percent for DEN-2, and 70 to 80 percent for DEN-3 and DEN-4. It will take more clinical experience before we'll know ultimately how effective the vaccine will be, especially against severe dengue hemorrhagic fever. We would call these results encouraging but still a work in progress.

In the meantime, five other dengue vaccine candidates are in development. But the timeline here makes an important point about public health: You can't just snap your fingers, throw a bunch of money at a problem, and expect an instant solution. The optimum scenario is to start developing solutions *before* the problem gets out of hand.

And we must always anticipate running into problems.

When dengue vaccines were first considered, concerns were raised that the antibody produced by vaccination might lead to an immune enhancement situation for those who are then exposed to the virus several years after vaccination, making them more vulnerable to DHF. In the summer of 2016, Dr. Scott

Halstead, one of the leading figures in dengue research of the past fifty years, sounded the alarm that recipients of CYD-TDV vaccine under five years old were five to seven times more likely to be hospitalized for severe dengue infection than those who did not receive the vaccine.

It is unclear what the data mean at this time, but they raise a lot of questions, such as whether the effect was limited to young children and whether the risk continues to increase over time after vaccination. Unless and until we figure it out, this is a real red-alert concern for this vaccine and all others in development.

Since the end of effective mosquito control in the 1970s, the home base for *Aedes* has expanded dramatically. A recent study estimates that today more than 3.9 billion people, in 128 countries, are at risk of infection with dengue viruses. This means they are also at risk for *Ae. aegyti*'s other afflictions: yellow fever, chikungunya, and Zika. There are a number of additional mosquito-borne viruses that one day may become the next *Aedes*-transmitted public health crisis, including Sepik, Ross River, Spondweni, and Rift Valley fever viruses. Like the Zika or chikungunya viruses were just a few years ago, these are problems no one has heard of yet.

In the past forty years, Gubler tells us, eradication efforts have been failures. In that time, there have been only two real successes in controlling *Ae. aegypti:* one in Singapore from 1973 through 1989, the other in Cuba from 1982 through 1997. Both of these campaigns ultimately failed, but for unrelated reasons. Singapore experienced a surge of economic growth, requiring the importation of hundreds of thousands of migrant workers, many from areas in which dengue was endemic. This factor, combined with an influx of tourists, substantially reduced the herd immunity. For Cuba, the problem came when the crumbling Soviet Union could no longer provide substantial financial aid. The *Ae. aegypti* program was one of the casualties. Both remind us that public health is inextricably bound up with every other societal factor.

Chikungunya

The word "chikungunya" is believed to have come from the Makonde language, spoken in southeastern Tanzania and northern Mozambique, and means "bent up," a pretty accurate description since one of the main symptoms of this *Aedes*-transmitted alphavirus is often severe joint pain. Other symptoms include fever, rash, fatigue, headache, conjunctivitis, and digestive tract distress. The mortality rate is low—less than 1 in 1,000—but the joint pain can last for months or years and may become a cause of chronic pain and disability.

Chikungunya was first isolated in Africa and by the 1950s it had spread to Asia, causing small epidemics in India, Myanmar, Thailand, and Indonesia. It seemed to disappear in the 1980s but reemerged in 2004 in East Africa. The new strain was highly transmissible, and within two years, India had about 1.3 million cases.

The first introduction of chikungunya virus into the Americas occurred in Saint Martin in late November 2014. We had a family vacation to Saint Martin planned for the following March. I realized with the confirmation of chikungunya cases on the island that it would spread quickly among the residents and visitors. Despite pushback from friends and family and complaints that I was overreacting, I canceled the condo reservation ninety-one days before our scheduled arrival (our contract provided for a full refund if we canceled more than ninety days prior). By the March week we had planned to be there, chikungunya virus transmission was in full swing on Saint Martin. By June 2016 it had spread to forty-five countries in the hemisphere, with more than 1.7 million cases and 275 deaths reported.

Though not a pleasant prospect to suffer through, we didn't regard chikungunya with the same sense of seriousness and urgency as we did some of the others. Yellow fever and dengue hemorrhagic fever could kill you, while chikungunya would most likely just make you miserable for a time. But now that the virus has settled in the Americas, we're learning that it might be more serious than we traditionally thought.

All of these viruses share *Ae. aegypti* as the primary vector. Its rural cousin *Ae. albopictus*—aka the Asian tiger mosquito—is starting to adapt to some of its habits and habitats and has become a secondary vector.

There is no magic bullet in controlling *Ae. aegypti* and *Ae. albopictus.* Studies have confirmed our conviction that good vector control is a complex science involving not only eliminating adult mosquitoes, but also reducing sources and using larvicide. We've also noted that there has been no development of a new, safe, and effective insecticide to replace DDT.

Today, no one public health organization or government agency is in charge of mosquito control. Imagine O'Hare Airport functioning without an air traffic control tower. That is what we have for global, regional, national, and even local *Aedes* control in the twenty-first-century world.

What we need is comprehensive, integrated, country-by-country mosquito control programs that target breeding-site elimination and, when that is not possible, breeding-site reduction. We need new and better tools to attack the adult mosquitoes, including new, effective pesticides and modern technologies such as genetically modifying mosquitoes. Finally, what we need are safe and effective human vaccines for the *Aedes*-transmitted viruses.

With the residual suspicion of DDT and the resistance built up by mosquitoes over the decades, new classes of insecticides will have to be developed that will provide at least six months of effectiveness in most climates. In continuously warm areas, spraying might have to take place at more than yearly intervals. It will be necessary to target both adult and larval mosquitoes.

Several attempts to enlist the mosquitoes in their own population control appear promising. Releasing sterile males into the *Aedes* population as eggs may decrease numbers in the wild. Field trials are being conducted in Malaysia, the Cayman Islands, Brazil, and Panama. I'm skeptical of this method of control due to the behavioral characteristics of *Aedes*. They typically won't fly more than several hundred feet from where they hatch, not even venturing across a road. In order for the sterile-male approach to work, mosquitoes would have to be distributed every hundred yards across the Americas. That would kind of be like building a ladder to the moon. It might be helpful on a limited local level, but it can't be the foundation of a national control program.

Another approach is to infect mosquitoes with *Wolbachia,* a common bacterium that interferes with virus transmission by the mosquito. A third involves genetically altering males so that eggs laid by the females never grow into mature bugs. A fourth experimental technique, known as gene drive, might be able to alter the mosquitoes' immune systems so that they block transmission of the virus.

While Gubler would like to see effective and safe vaccines developed that could be adapted to all or some of the *Aedes*-transmitted arboviruses, he warns that this, in itself, will never be a successful solution on its own. He believes, and I strongly agree, that a rigorous, integrated approach involving a paramilitary-style program of spraying, effective mosquito bed nets in vulnerable areas without air conditioning or window screens, and genetic manipulation and control of mosquito populations all must be instituted together to achieve any significant and lasting progress against *Ae. aegypti* and its related species. As we've seen with so many other diseases, poor countries in the developing world may not have the means to buy drugs and vaccines and will have to rely on their own resources.

In light of how fragmented the leadership in vector-borne disease control is, on a global, regional, national, and local level, Gubler and a coalition of his expert colleagues have proposed the creation of a global alliance of institutions that have a vested interest in preventing *Aedes*-transmitted diseases. The proposed name is the Global Alliance for Control of *Aedes*-Transmitted Diseases (GAAD), and it would include NGOs, international funding agencies, and foundations. Its operational arm, known as the Global Dengue and *Aedes*-Transmitted Diseases Consortium (GDAC), is intended to work closely with the WHO and selected international and governmental organizations.

My continual complaint when I don't see important and rational steps being taken against major disease threats is, "No one is in charge!" So when we see a group of responsible experts ready and willing to take a leadership role, my first—and lasting—instinct is to pledge my enthusiastic support.

CHAPTER 15

Zika: Expecting the Unexpected

The rapidly evolving outbreak of Zika warns us that an old disease that slumbered for 6 decades in Africa and Asia can suddenly wake up on a new continent to cause a global health emergency.

— WHO Director-General Margaret Chan, MD, May 23, 2016

An infectious disease that had been known for almost seventy years suddenly became a household word when Zika virus appeared in much of the Western Hemisphere in the spring of 2016. Everyone seemed shocked that this new infection that caused horrifying birth defects had seemingly appeared out of nowhere. But Zika didn't just appear out of nowhere to begin its course in the Americas. Many of my colleagues were just not paying attention to what Mother Nature was in the process of doing. They weren't looking in the right place.

Zika was first detected in a rhesus monkey in the Zika Forest in Uganda in 1947 and then was isolated in a ten-year-old girl in Nigeria in 1954. Its first Asian sighting was in 1966, when the virus was isolated from *Ae. aegypti* in Malaysia. Compared to the really bad stuff, like malaria and yellow fever, the symptoms of Zika appeared mild—conjunctivitis, a pink rash, and sometimes joint and muscle pain, or no symptoms at all. For fifty years, there were no more than twenty documented cases of human disease, and most of those were picked up incidentally in tests for yellow fever. No one even thought about working on a vaccine.

Public health officials watched with interest but little alarm as the Zika virus made its way across the Pacific, to the island of Yap in Micronesia in 2007. By 2013 it had reached French Polynesia, and that is where the international public health monitors should have picked it up and realized something scary was happening.

Between October 2013 and February 2015, 262 Zika virus infections were documented there. Among those cases were seventy individuals who presented with neurological or autoimmune complications, including thirty-eight cases of Guillain-Barré syndrome (GBS).

Guillain-Barré, sometimes called French polio, is caused by an autoimmune reaction: An antibody attacks the myelin sheath—the coating covering the body's nerves. When the coating is damaged, the nerve can't maintain its electrical conduction. About half the cases occur shortly after an infection. Common causes are *Campylobacter* bacteria, cytomegalovirus, and Epstein-Barr virus.

Some cases are extremely mild. Others can be frightening and require hospitalization. GBS is normally transitory, as the myelin sheath grows back. This can take anywhere from several weeks to months. However, in the meantime, it often requires intensive treatment, and for those in frail health to begin with, or in particularly severe cases in previously healthy individuals, it can affect the breathing muscles and can lead to death. Even with First World treatment, about 10 percent of victims will suffer lasting effects. In areas of the developing world where good medical support is not available, GBS is likely to result in more fatalities and lasting effects.

The fact that certain viral and bacterial infections can, in relatively rare cases, trigger GBS was not a new finding, and infectious disease specialists are always on guard against it in their seriously ill patients. But nothing this severe had previously been noted with Zika, and when they noted GBS, the French Polynesian medical community regarded the virus with growing alarm.

One group that did have their public health eyes focused on the French Polynesia Zika outbreak was the European Centre for Disease Prevention and Control (ECDC). They published a comprehensive rapid risk assessment on the situation on February 14, 2014. While it wasn't completely clear if somehow dengue virus infection, together with Zika, was responsible for this new clinical spectrum, it was definitely a concern. I remember reading the ECDC report and thinking that, since *Ae. aegypti,* and possibly *Ae. albopictus,* were responsible for the Zika virus transmission in French Polynesia, we had all the ingredients we needed for it to take off in the Americas.

The year after it struck French Polynesia, Zika spread to New Caledonia and the Cook Islands, hopping from island to island until it reached Easter Island, its gateway to the Americas: all completely predictable.

While we never should have been surprised by Zika's arrival on our doorstep, we could not have known the full extent of the danger. The French Polynesia outbreak did not give us an early clue that microcephaly would be such a serious complication of Zika infection. Those data came later. The 2016 version of Zika virus turns out to be a lot more serious than even I thought possible.

By the early months of 2015, doctors in cities along the east-central coast of Brazil were seeing a dramatic rise in GBS cases, with patients often noticing a rash on their bodies a few days before diagnosis. By summer, the really bad news hit. An increasing number of babies were being born with microcephaly: a birth defect where a baby's head is smaller than normal and the brain does not develop properly. Often the new mothers reported experiencing a rash during their pregnancy, particularly in the first trimester. This condition is independent of GBS.

Because of the spike in such births, Brazilian physicians and medical scientists quickly suspected a connection between Zika and microcephaly. This is absolutely devastating for any parent, and in Brazil it was compounded by the fact that so many of the births were in abjectly poor families with little or no outside support. It turns out that the Zika virus directly invades the fetal nervous system during pregnancy. The head CT of a normal baby compared to a head CT of a baby with microcephaly shows evident and frightening differences. In the afflicted baby, there is more space between the brain and the skull as well as unusual dark regions within the brain itself.

By mid-January 2016, the CDC issued recommendations warning pregnant women of the risk of Zika-related complications and the role that sexual transmission could play in new infections. Despite the rapidly growing body of data supporting the cause-and-effect relationship role of Zika virus and microcephaly and GBS, many of my academic-based infectious disease colleagues and the news media were slow to arrive at that same conclusion. During January and February 2016, Zika reporting often revolved around a debate about whether the virus caused microcephaly and GBS.

For me, this discussion seemed like such a waste of time, like two firefighters arguing about who gets to drive the truck to the burning building. For those of us who have spent our careers on the front lines of outbreak response, there was no doubt that Zika virus was causing an ever-growing number of adverse health outcomes.

This issue came to a head for me during the last weekend of January 2016, when the *New York Times* asked me to write a Sunday op-ed on what we should know about the emergence of Zika. I stated plainly that it caused microcephaly and GBS. The editors in charge of my piece got back to me the Friday afternoon before its publication and informed me I wouldn't be allowed to say that in the article because the *Times*' health reporting team had not yet reached a similar conclusion.

I didn't care what the *Times*' health reporters had concluded; Zika *was* causing these conditions. After several calls lasting more than an hour and no successful resolution of this point, I requested that my op-ed be pulled. I was not going to publish something that would add to the unnecessary confusion about the emerging Zika crisis just to get another *New York Times* op-ed byline. Finally the powers that be at the *Times* decided to allow the statement. Our job now was to stop this silly debate and get on with doing whatever we could to minimize its impact, and I said so in my op-ed piece.

Today we know that microcephaly and a growing number of other birth defects, including craniofacial disproportion, spasticity, seizures, irritability, eye problems, and brain-stem dysfunction result from Zika virus infection during pregnancy. Recent studies by the CDC and Brazilian researchers found that between 1 and 13 percent of women infected during the first trimester of pregnancy deliver babies with microcephaly.

By the time the connection showing that Zika caused GBS and microcephaly was confirmed less than a year after its arrival in the Americas, the virus had taken on the persona of a twenty-first-century thalidomide tragedy; thalidomide was the German sedative and morning sickness antidote of the late 1950s and early 1960s that led to babies born with missing, short, or flipper-like limbs, vision and hearing problems, and deformed hearts and other organs. For decades, the mere mention of thalidomide struck fear into the hearts of pregnant women. Now the same was happening with Zika. The difference was that with thalidomide, you had to actively take a pill to risk these birth defects. With Zika, you just had to passively be bitten by an *Aedes* mosquito. And mosquitoes were all around.

Seldom has an infectious disease actually led to the recommendation not to get pregnant, even though we know of two others that can cause heartbreaking birth defects.

The first, congenital rubella syndrome, can occur in a fetus whose mother has been infected with rubella (German measles) during pregnancy, and the risk is highest during the first twelve weeks of gestation. Hearing impairment is the most common result, but there can also be eye problems such as cataracts, congenital heart disease, and developmental problems. A vaccine was licensed in the United States, where rubella has essentially been eliminated, but it continues to be endemic in many parts of the world. The CDC estimates that more than 100,000 babies are born with congenital rubella syndrome every year.

Second, here in the United States about 30,000 children are born each year with congenital cytomegalovirus, a common virus that rarely produces symptoms but that can be serious for anyone with a weak immune system and for pregnant women. In the latter, it can produce low birth weight in the baby, jaundice, enlarged spleen and enlarged, poorly functioning liver, pneumonia, and seizures. So far, there is no treatment.

As tragic as these two conditions can be, the worst-case Zika scenario is an order of magnitude higher.

One of the most dramatic aspects of the Zika epidemic is the frequency of sexual transmission of the virus. Though other flavivirus infections such as dengue and yellow fever have been studied extensively in humans for more than a hundred years, sexual transmission was never documented. We now have to fight infection from multiple human "ports of entry." A mosquito bite, sexual intercourse, or blood transfusion will transmit the Zika virus efficiently. There is even limited evidence that caregivers can become infected through contact with the body fluids of patients with Zika virus.

Brazilian researchers recently found that women in the sexually active age-group are overwhelmingly more likely than men to be infected with Zika virus, with sexual transmission the most likely cause. This may be due to the relative efficiency of male transmission to female rather than vice versa. It may also be due to a greater number of women than men seeking testing because of the pregnancy risks.

Infection of pregnant women has led to a series of tough public health and policy issues, including the availability and use of contraception in the largely Catholic countries of the Americas, abortion of fetuses shown by imaging to have microcephaly, and the recommendation that women of childbearing years delay pregnancy if possible. Based on our previous experience with the introduction of a new mosquito-borne flavivirus into a population with no previous history of infection, there tends to be dynamic transmission and lots of cases for three to four years. After that time, a high percentage of the population will have been infected and develop immunity. It's likely that the risk for getting infected with Zika will be substantially lower in the Americas in 2020 than it is in 2016. But making the recommendation to delay pregnancy has been extremely controversial in the Zika outbreak.

As of August 1, 2016, the CDC was reporting 1,825 confirmed cases in forty-six of the fifty United States, 479 of which were pregnant women. Sixteen of those cases were sexually transmitted and 5 led to GBS. There were an additional 5,548 cases in US territories, of which 493 were pregnant women and 18 cases contracted GBS. This, of course, is only the beginning. A recent CDC study documented that an estimated 216.3 million passengers travel annually by air, sea, or land to the United States from areas with local Zika virus transmission. In addition, an estimated 51.7 million passengers are women of childbearing age and 2.3 million are pregnant at their time of arrival in the United States.

Previously, all cases were acquired either outside the continental United States or as the result of sexual transmission from someone who had traveled from a high-risk area. But as of August, there was evidence of mosquito-borne transmission within a localized area of Miami-Dade County. It is likely that similar transmission will occur in other areas of the Gulf Coast.

Zika has already done serious damage to tourism in the Caribbean and has now moved into Florida. During House and Senate debates in spring 2016 on funding for Zika prevention, Marco Rubio, the Republican senator from Florida, sided with the Democrats in urging that new money be approved. "There is just a lack of urgency about it," he told the *New York Times*. "People are going to be asking, 'Why didn't you do anything?' You are going to have to have a pretty good answer, and I am not sure there is going to be one."

Being a Floridian, Rubio knows his state could take a severe hit: "I tell people we are just one mosquito-borne infection away from serious damage to our tourist industry."

As BARDA former acting director Richard Hatchett, MD, said, "Ebola was easy to contain—until it wasn't. The same could be true with Zika."

The first questions that confront us in the public health community are: Why did Zika become so much more dangerous so quickly? Was it always that way and we just didn't have a large enough patient cohort to realize it? Or had something changed?

Duane Gubler thinks it comes down to mutation. "We know that mutations or small genetic changes can dramatically affect the epidemic potential, and probably virulence, of dengue and chikungunya viruses," he says, "so probably Zika, too."

Gubler thinks the jump in raw numbers triggered by epidemic spread of Zika on its own could be causing the increase in birth defects and more serious symptoms. But it is most likely that a change in the actual genetics of the virus is the greatest contributor, an analysis I find completely reasonable. Time and more research will clarify if this is the reason for the sudden change in the epidemiology of Zika infection. Nonetheless, Zika is a humbling reminder that the current epidemiology of a human infectious disease, in particular one caused by viruses, can change at any time. I'm certain we are in for more surprises.

There is no treatment for Zika virus other than supportive care in hospitals, and there are no effective preventative drugs or antivirals. While at least twelve pharmaceutical companies, universities, and government agencies have expressed interest in pursuing effective and safe Zika vaccines, they won't be available anytime soon.

Keeping in mind the antibody-dependent enhancement we discussed previously with dengue vaccine, I'm certain that no regulatory agency, such as the Food and Drug Administration, will license a Zika vaccine without an abundance of safety data. This could mean vaccinating and following many thousands of study participants. So even if a safe and effective Zika vaccine is possible, it is still years off.

If the virus that has exploded in the Americas is, indeed, a recently mutated and more dangerous pathogen, it remains to be seen if infection with a previous version of the virus will confer protection against this strain. We have no idea how many people in Asia and Africa are currently protected against the current virus.

There are forty-two countries and territories in the Americas that have confirmed local mosquito-borne transmission of Zika virus. The possibility that we could see similar outbreaks in Africa and Asia has to be factored into any consideration of the problem. Remember from the previous chapter that there are an estimated 3.9 billion people in 128 countries who are at risk of infection with dengue viruses. The same number must be considered at risk for Zika.

Zika is the first public health crisis of my career that has come down to a partisan battle over needed resources. This bodes poorly for future crises and raises serious questions about our ability to respond to future challenges.

Throughout the summer of 2016, news cameras focused on shots of government spraying programs. This might have made viewers feel good, but the spraying offered little real protection. Spraying does not kill the mosquito larvae and cannot reach all the areas, indoors and out, where *Aedes* breeds and resides.

Duane Gubler has expertise in this area. In 1987, he conducted a study on spraying in Puerto Rico during a large dengue fever outbreak using the same type of airplanes and the same insecticide, called naled. He found that while spraying was effective in decreasing mosquito populations, it did nothing to reduce the transmission of dengue.

Zika and all the other *Aedes*-transmitted diseases are going to be a trench-warfare slog against the mosquito and the viruses it carries, using every means we have, while we attempt to develop new and more effective ways of combating them.

In the meantime, continue to expect the unexpected.

CHAPTER 16

Antimicrobials: The Tragedy of the Commons

The thoughtless person playing with penicillin treatment is morally responsible for the death of the man who finally succumbs to infection with the penicillin-resistant organism. I hope this evil can be averted.

— Sir Alexander Fleming, MD

About 4 million years ago, a cave was forming in the Delaware Basin of what is now Carlsbad Caverns National Park in New Mexico. From that time on, Lechuguilla Cave remained untouched by humans or animals until its discovery in 1986—an isolated, pristine primeval ecosystem.

An article by Dr. Kirandeep Bhullar of Ontario's McMaster University and seven others, published in the April 2012 issue of the peer-reviewed online journal *PLoS One,* received little notice outside the scientific community. But its implications were provocative and sobering.

When the bacteria found on the walls of Lechuguilla were analyzed by the article's authors, many of the microbes were determined to have resistance not only to natural antibiotics like penicillin, but also to synthetic antibiotics that did not exist on earth until the second half of the twentieth century. As infectious disease specialist Brad Spellberg, MD, put it in the *New England Journal of Medicine,* "These results underscore a critical reality: antibiotic resistance already exists, widely disseminated in nature, to drugs we have not yet invented."

The origin story of antibiotics is well-known, almost mythic: Returning to his lab at St. Mary's Hospital in London in 1928 after a holiday, Dr. Alexander Fleming noticed that a fungus had corrupted one of his staphylococci culture petri dishes and that the staph colonies surrounding the fungus had been destroyed. This was every bit the equal of the observation that English milkmaids didn't get smallpox.

Fleming grew this fungal mold in a pure culture and found that the result killed a range of disease-causing bacteria. The mold was from the *Penicillium* genus, so he called it penicillin. It was left to Drs. Howard Florey and Ernst Chain to figure out penicillin's structure and transform it into a lifesaving medical agent. The three pioneers shared the Nobel Prize in Physiology or Medicine in 1945.

At around the same time that Florey and Chain were working in England, a team at a division of IG Farben in Germany (later to become Bayer) led by Dr. Gerhard Domagk was exploring the properties of red chemical dyes called sulfonamides: substances derived from coal tar that did not kill bacteria but inhibited their growth. They became the basis for a group of medicines known as sulfa drugs, the first of which was marketed as prontosil. In 1933, one of Domagk's colleagues treated a ten-month-old baby boy with an almost always fatal *S. aureus* infection in his blood. The boy became the first person in history whose life was saved by an antimicrobial.

Ironically, two years later, Domagk's six-year-old daughter lay near death from a massive infection after accidentally puncturing her hand with a sewing needle. Her doctor recommended amputating the arm in a desperate attempt to stem the infection. Instead, just as desperately, Domagk administered prontosil. Within four days, the little girl had recovered. Domagk was awarded a Nobel Prize in 1939.

Nor did it stop there, so great was this medical revolution. Dr. Selman Waksman, the Russian-born American biochemist and microbiologist who suggested the use of the term "antibiotic," was awarded the Nobel Prize in 1952 for the discovery of streptomycin—purified from soil bacteria—the first such agent that could treat tuberculosis.

Today, heart disease and cancer are, by far, the leading causes of death in the United States. In 1900, they were relatively insignificant. This is not because our forebears pursued a healthier lifestyle, didn't smoke, or followed a more prudent diet. It's because back then infectious diseases didn't give our two modern killers a chance to move in; they got to people earlier and more often than heart disease and cancer ever could. Antibiotics, along with the other basic public health measures we have described, have had a dramatic impact on the quality and longevity of our modern life. When ordinary people called penicillin and sulfa drugs miraculous, they were not exaggerating. The discoveries of Domagk, Fleming, Florey, and Chain ushered in the age of antibiotics, and medical science assumed a lifesaving capability previously unknown.

Note that we use the word "discoveries" rather than "inventions." Antibiotics were around many millions of years before we were. Since the beginning of time, microbes have been competing with other microbes for nutrients and a place to call home. Under this evolutionary stress, beneficial mutations occurred in the "lucky" and successful ones that resulted in the production of chemicals—antibiotics—to inhibit other species of microbes from thriving and reproducing, while not compromising their own survival. Antibiotics are, in fact, a natural resource—or perhaps more accurately, a natural phenomenon—that can be cherished or squandered like any other gift of nature, such as clean and adequate supplies of water and air.

Equally natural, as Lechuguilla Cave reminds us, is the phenomenon of antibiotic resistance. Microbes move in the direction of resistance in order to survive. And that movement, increasingly, threatens our survival.

The World Economic Forum's *Global Risks 2013* report declared, "While viruses may capture more headlines, arguably the greatest risk of hubris to human health comes in the form of antibiotic-resistant bacteria. We live in a bacterial world where we will never be able to stay ahead of the mutation curve. A test of our resilience is how far behind the curve we allow ourselves to fall."

In his book *Missing Microbes,* Dr. Martin Blaser explains how our use of antibiotics over the past eighty years is greatly altering the three-billion-year-old microbiome that resides in our bodies. He lays out with clarity and vision why what I call "supermicrobial evolution in our modern world" poses a real and new danger for our future encounters with infectious diseases. What we are dealing with, to put it plainly, is a slow-motion worldwide pandemic. With each passing year, we lose a percentage of our antibiotic firepower. In a very real sense, we confront the possibility of revisiting the dark age where many infections we now consider routine could cause severe illness, when pneumonia or a stomach bug could be a death sentence, when a leading cause of mortality in the United States was tuberculosis. The most comprehensive and accurate assessment of the future of antimicrobial resistance and the devastating impact it will have on humans and animals in the years to come is the *Review on Antimicrobial Resistance,* a detailed study commissioned by the British government of Prime Minister David Cameron and supported by my friends and colleagues at the Wellcome Trust. (Cameron reaffirmed the seriousness he places on this issue when he mentioned it on April 22, 2016, during a joint news conference with President Obama in London, as part of his enumeration of the top challenges facing the modern world.) The effort became known as AMR and was led by Lord Jim O'Neill, an internationally recognized macroeconomist, former chairman of Goldman Sachs Asset Management, and former British government minister.

Many people wondered why an economist was chosen to chair such an important medical study. But I believe he was the perfect choice, because every aspect of this problem is tied to economic issues—for governments, for the pharmaceutical industry, for world agriculture, and for the practice of healthcare, much of which is paid for through reimbursements. Macroeconomists are trained to look at the big picture. O'Neill is one of the world's best macroeconomists. He is the man who coined the acronym BRIC for Brazil, Russia, India, and China and who has a firm understanding of what role those nations must play in the critical effort against antimicrobial resistance.

After studying the issues for more than two years, O'Neill and his highly talented team of researchers determined that, left unchecked, in the next thirty-five years antimicrobial resistance could kill 300 million people worldwide and stunt global economic output by $100 trillion. There are no other diseases we currently know of except pandemic influenza that could make that claim. In fact, if the current trend is not altered, antimicrobial resistance could become the world's single greatest killer, surpassing heart disease or cancer.

The problem of drug resistance isn't new. Dr. Max Finland, a world renowned professor at Harvard Medical School and a pioneer in the development and use of antibiotics for almost fifty years, convened eight international experts on infectious diseases in 1965 and asked the question "Are new antibiotics needed?" The results of that conference were published in a major medical research journal later that year. The conclusion reached by the group was a resounding *yes:* We need new antibiotics to cover diseases not yet well treated and because of the diminishing effectiveness of antibiotics available due to the emergence of antibiotic resistance. Our current discussions, therefore, are like déjà vu all over again.

The only difference between then and now is that whole fleets of antibiotics that were available in 1965 or discovered after that time are now additional clinical casualties of antibiotic resistance. The rate of that resistance now far exceeds the rate of new antibiotic development. In some parts of the United States, about 40 percent of the strains of *Streptococcus pneumoniae,* which the legendary nineteenth- and early-twentieth-century physician Sir William Osler called "the captain of the men of death," are now resistant to penicillin. And the economic incentives for pharmaceutical companies to develop new antibiotics are not much brighter than those for developing new vaccines. Like vaccines, they are used only occasionally, not every day; they have to compete with older, extremely cheap generic versions manufactured overseas; and to remain effective, their use has to be restricted rather than promoted.

As it is, according to the CDC, each year in the United States at least 2 million people become infected with antibiotic-resistant bacteria and at least 23,000 people die as a direct result of these infections. More people die each year in this country from MRSA (methicillin-resistant *S. aureus,* often picked up in hospitals) than from AIDS.

Most of us can't quite imagine that time before Domagk, Fleming, Florey, and Chain, in which our great-grandparents and, in some cases, even our grandparents lived, before the antibiotic era that has been our great gift since the late 1940s. But within ten to twenty years, we could well be moving into the postantibiotic era.

If we can't—or don't—stop the march of resistance and come out into the sunlight, what will a postantibiotic era look like? What will it actually mean to return to the darkness of the cave?

Well, for one thing, clearly, more people will get sick and more people will die from germs we've been able to combat for the past seventy years. But once we get down in the weeds, it's even more chilling. Without effective and nontoxic antibiotics to control infection, any surgery becomes inherently dangerous, so all but the most critical, lifesaving procedures would be complex risk-benefit decisions. You'd have a hard time doing open-heart surgery, organ transplants, or joint replacements, and there would be no more in vitro fertilization. Caesarian delivery would be far more risky. Cancer chemotherapy would take a giant step backward, as would neonatal and regular intensive care. For that matter, no one would go into a hospital unless they absolutely had to because of all the germs on floors and other surfaces and floating around in the air. Rheumatic fever would have lifelong consequences. TB sanitaria could be back in business. You could just about do a postapocalyptic sci-fi movie on the subject.

How did we get here? To understand why antibiotic resistance is rapidly increasing and what we need to do to avert this bleak future and reduce its impact, we have to understand the big picture of how it happens, where it happens, and what the major drivers are.

They are, in ascending order of magnitude:

1. Human use in the United States, the United Kingdom, Canada, and the European Union—the countries that have done the most to foster antibiotic stewardship, though many challenges remain.
2. Human use in the rest of the world, where little has been done to curb resistance to date.
3. Use for animals in the United States, Canada, and Europe, where the food livestock, poultry, and fish industries have been largely unwilling to address the issue of overuse without serious pressure from government and the public health sector.
4. Use for animals in the rest of the world, which we don't have reliable data on, but which we know is high and increasing.

Let's take a look at each of our four categories of resistance by human and animal demographics and geography.

Human Use in the United States, the United Kingdom, Canada, and the European Union

Think of an American couple, both of whom work full-time. One day, their four-year-old son wakes up crying with an earache. Either mom or dad takes the child to the pediatrician, who has probably seen a raft of these earaches lately and is pretty sure it's a viral infection. They almost always are. There is no effective antiviral drug available to treat the ear infection. Using an antibiotic in this situation only exposes other bacteria that the child may be carrying to the drug and increases the likelihood that an antibiotic-resistant strain of bacteria will win the evolutionary lottery. But the parent knows that unless the child has been given a prescription for *something,* the daycare center isn't going to take him, and neither partner can take off from work. This is a real everyday problem, and it doesn't seem like a big deal to write an antibiotic prescription to solve this couple's dilemma, even if the odds that the antibiotic is really called for are minute.

But it is a classic "Tragedy of the Commons." As Spellberg explained in his pioneering 2009 book, *Rising Plague:*

> First described by Garrett Hardin in *Science* magazine in 1968, the "Tragedy of the Commons" applies to scenarios where an individual acts to significantly benefit [himself], and as a consequence accepts as a tradeoff a small amount of overall harm to society at large. If only one person is so acting, the total harm to society is small. But when everyone in society undertakes the same action, the collective harm to everyone becomes enormous.

Several surveys show that while the majority of people understand that antibiotics are overprescribed and therefore subject to mounting resistance, they think the resistance applies to *them,* rather than the microbes. They believe that if they take too many antibiotics—whatever that unknown number might be—they will become resistant to the agents, so if they are promoting a risk factor, it is only for themselves rather than for the entire community.

Doctors, of course, understand the real risk. Are they culpable to the charge of over- and inappropriately prescribing antibiotics? In too many cases, the answer is *yes.*

In the May 3, 2016, issue of the *Journal of the American Medical Association,* the CDC published the results of a study undertaken with the Pew Charitable Trusts and other public health and medical experts. The study found that in physicians' offices and hospital emergency departments, at least 30 percent of antibiotic prescriptions are unnecessary or inappropriate. Not surprisingly, most are given for respiratory conditions such as colds, sore throats, bronchitis, and sinus and ear infections that are caused by viruses.

The CDC's press release states, "These 47 million excess prescriptions each year put patients at needless risk for allergic reactions or the sometimes deadly diarrhea, *Clostridium difficile.*" This brings up another important point. Not only does overuse accelerate antibiotic resistance, but these agents are not completely benign. Like many drugs that treat serious conditions, they have side effects—in the CDC's example, by possibly wiping out the "good" and necessary bacteria in the gut.

Why do doctors overprescribe? Is it about covering their backsides in this litigious society? Is it a lack of awareness of the problem? According to Spellberg, "The majority of the problem really revolves around *fear.* It's not any more complicated than that. It's brain-stem-level, sub-telencephalic, not-conscious-thought fear of being wrong. Because we don't know what our patients have when they're first in front of us. We really cannot distinguish viral from bacterial infections. We just can't.

"You can say on a population basis that 95 percent of patients who present with these signs and symptoms have a virus. But when I have an individual in front of me and I'm going to see 10,000 of these individuals in my career, I'm going to be wrong sometimes. And if I'm wrong, the consequences could be really bad. That's what drives most of it. And patients suffer from the same fear. They come, they don't feel well, they want something. They don't want to get into a philosophical debate. They want something that's going to make them feel better. That's why they ask for the prescription."

Spellberg cited a couple of cases for us. In the first, he got a call from a chief resident in surgery, saying she had a patient with an infected gallbladder. The patient was taking the correct, fairly narrow-spectrum antibiotic—one that targets a limited number of bacteria—but her white blood cell count was going up (a sign of the body's response to infection), her fever was continuing to rise, and the pain was getting worse. So the resident wanted to put the patient on piperacillin-tazobactam, known commercially as Zosyn—a powerful broad-spectrum antibiotic that kills *Pseudomonas aeruginosa,* one of the worst pathogens out there.

Spellberg asked why she would want to use that particularly valuable antibiotic when there was virtually no chance the patient had *Pseudomonas.* The resident explained that she wasn't worried about *Pseudomonas,* but the patient was continuing to get worse.

"Yeah," he replied. "But the patient's getting worse because you need to take out her gallbladder."

"Well," she said, "there were a couple of trauma cases that bumped her from the OR so we couldn't operate right away, and I just want to broaden the antibiotic."

"This is completely irrational," Spellberg says. "And the resident knows it's irrational, but *she's afraid.* She wants the Band-Aid of broad-spectrum antibiotics to make *herself* feel better."

In the next case, he got a request from a resident for Cipro, another powerful broad-spectrum antibiotic, for a patient with gram-negative bacteria in her urine. Gram-negative is one of the two main classifications of bacteria, characterized by their type of cell membrane and identified by not reacting to a special lab stain. Gram-positive, not surprisingly, is the other type. They are named for the inventor of the staining technique, Danish bacteriologist Hans Christian Gram.

Spellberg asked what the patient's symptoms were and was told there weren't any. "So the question is: How do we treat asymptomatic bacteriuria [bacteria in the urine]? And the answer is: We don't. This is cognitive dissonance staring us in the face. If this resident had this question on a board exam, he'd get it correct. But that's a piece of paper and this is a patient staring him in the face, and he's afraid. And we have not tackled the fear. We've got to figure out psychological ways of getting around the fear."

Now, after hearing these two cases, you wouldn't be out of line for thinking that doctors, particularly young doctors, just have to get it together and start thinking critically and rationally about each case. Then Spellberg throws one more case at us, one he heard at an infectious disease conference he attended:

A twenty-five-year-old woman came into the urgent care facility of a prominent healthcare network complaining of fever, sore throat, headache, runny nose, and malaise. These are the symptoms of a classic viral syndrome and the facility followed exactly the proper procedure. They didn't prescribe an antibiotic, but instead told her to go home, rest, keep herself hydrated, maybe have some chicken soup, and they would call her in three days to make sure she was all right.

She came back a week later in septic shock and died soon after.

"It turns out she had Lemierre's syndrome," says Spellberg. "It clotted her jugular vein from a bacterial infection that spread from her throat to her bloodstream. This is about a 1-in-10,000 event; it's pretty darn rare. But it's a complication of an antecedent viral infection, and it's a known complication. So this patient, ironically, would have benefited from receiving inappropriate antibiotics."

Mark's brother Jonathan Olshaker, MD, is chief of the Emergency Department at Boston Medical Center, the largest safety-net hospital and busiest Level I trauma and emergency services center in New England. He is highly sensitive to the growing resistance problem, but also sensitive to doctors' and nurses' concerns about making mistakes that could hurt the patient.

"One thing no emergency physician wants to hear," Jon says, "is 'Remember that case you saw last week...?' Because you know the next line is going to be, 'Well, here's what happened to him...'"

"How many times do you think doctors need to have those things happen before they start giving antibiotics to every person who walks in the door?" asks Spellberg.

Human Use in the Rest of the World

The populations of the nations we have just discussed add up to about 868,798,000, or about 12 percent of world's population. Even if we make significant strides in reducing the rate of increase in antibiotic resistance evolution in this "First World," it will have only a short-term and limited impact on the eventual global catastrophe if we don't make this an international priority.

The BRIC countries are all at about the same level of development. Their combined population is around 3,938,300,000, or about 54 percent of the world's total. Then there is the rest of the planet; approximately 2,494,400,000 people, making up the remaining 34 percent. As much difficulty as we're having controlling antibiotic resistance in "our" 12 percent of the population, for the remaining 88 percent, we believe the situation to be a whole lot worse.

In many of these countries, antibiotics are sold right over the counter just like aspirin and nasal spray; you don't even need a doctor's prescription. While over-the-counter sales of antibiotics without prescriptions are illegal in numerous places around the world, lax enforcement results in extensive sales in many low- and middle-income countries.

While we in the public health community would certainly like to see a complete cessation of antibiotic use without a doctor's prescription, how do we tell sick people in developing countries that they first have to see a doctor, when there may be no more than one or two physicians for thousands of individuals, and even if they could find one, they couldn't afford the visit in the first place? Taking an action in a vacuum, such as banning over-the-counter sales without improving infrastructure, simply isn't viable.

We also have to understand the inordinate burden antibiotic resistance places on the world's poor. Current effective antibiotics now out of patent may cost only pennies a dose. When those are no longer useful, new compounds will cost many dollars a dose—far more than the poor can afford.

In an analysis commissioned by AMR, the London School of Economics found that in just four economically emerging nations on three continents—India, Indonesia, Nigeria, and Brazil—nearly 500 million cases of diarrhea are treated with antibiotics each year, a number expected to rise to more than 600 million by 2030. This gives us some sense of the scope of the problem, as well as underscoring the effects of unsafe water and unsanitary conditions. And what happens if the growing resistance problem means that at some point in the future, we can't treat these diarrheal cases with antibiotics affordable in the developing world?

Many of the antibiotic compounds in the developing world are produced in loosely regulated or unregulated manufacturing facilities, where there is no way to gauge quality control. And millions of poor people are living in tightly packed urban slums with inadequate hygiene and sanitary conditions, which generate both more disease and more opportunity for microbes to share resistance characteristics with one another.

To get some perspective on the challenge of resistance in the developing world, let's look at tuberculosis, one of the most devastating diseases of the nineteenth and early twentieth centuries. In various parts of the world, particularly Asia, tuberculosis has gone from being a disease largely treatable with antibiotics to a disease with some strains that are now labeled MDR (multidrug resistant), XDR (extensively drug resistant), or TDR (totally drug resistant).

And this is not just happening far from our shores. "I've been there for TB patients," states Dr. Tom Frieden, director of the CDC. "I've cared for patients in the US for whom there are no drugs left. It is a feeling of such horror and helplessness. This is not where we need to be." If we are confronted with this problem in the United States, imagine the challenges for the developing world.

Maryn McKenna, one of the leading independent journalists on public health and author of *Beating Back the Devil* and *Superbug,* tells us that "in various places in the US, anywhere with populations from areas around the world where these strains are seen, we are now having TB patients having pieces of their lungs removed. That's nineteenth-century medicine!" She has been studying antibiotic practice, policy, and resistance for more than a decade. So far, the problems have far outpaced the solutions.

Use for Animals in the United States, Canada, and Europe

But all of the world's use of antibiotics for humans is a relatively small percentage of *total* use. The United States, Canada, and Europe use about 30 percent of our antibiotics on humans. The rest we use on animals—specifically, animals we kill for food or companion animals.

We buy antibiotics for ourselves by the gram in little white or orange plastic bottles, sometimes in small blister packs. Industrial farmers and cattle ranchers buy antibiotics by the ton.

There are four applications for antibiotic use in raising food animals, all of which, to one extent or another, result from the way we go about protein-food production in the modern world. We produce our food animals in very large numbers and raise them densely packed together, whether we're talking about chicken and turkey operations, cattle and swine feedlots, or industrial fish farms. While these animals are less likely to catch infectious diseases when large production operations use high levels of biosecurity—the practice of limiting the ways that disease-causing germs can contact the animals—when these germs do get introduced, their spread is rapid and extensive. So we use antibiotics to treat the resulting infections. But we also use them to prevent infections in the first place, or to control them by dosing healthy animals so they don't catch anything from the sick ones. And then we use them to enhance growth.

In the late 1940s, fishermen near Lederle Laboratories in New York State noted that trout seemed to be larger than before. When Dr. Thomas Jukes, a prominent biochemist, investigated the apparent phenomenon with his colleague Dr. Robert Stokstad, they found that the antibiotic Aureomycin in the runoff from Lederle's plant was the cause. After experimentation with livestock and poultry produced similar results, the serendipitous discovery was hailed as an agricultural breakthrough.

For decades we have given food-production animals repeated doses of certain antibiotics to make them grow bigger and fatter, producing more meat per animal. This practice is known as growth promotion. The FDA has implemented a voluntary plan with the agriculture industry to phase out the use of certain antibiotics for growth promotion. The European Union banned this use in 1969, though they still use antibiotics for infection prophylaxis, control, and treatment. The AMR report found mounting evidence that the use of antibiotics for growth promotion may provide only very modest benefits to farmers in the high-income countries, usually less than 5 percent additional growth.

How does this antibiotic use affect us? The AMR team reviewed 280 published, peer-reviewed research articles that address the use of antibiotics in food production. Of these published studies, 139 came from research groups at academic institutions; 100, or 72 percent, found evidence of a link between antibiotic use in animals and antibiotic resistance in humans. Only seven articles, 5 percent, found no link between antibiotic use in animals and human infections.

In 2015, alarmed by reports of growing resistance, the Obama administration established the Presidential Advisory Council on Combating Antibiotic-Resistant Bacteria—PACCARB, since every government entity seems to get an acronym attached to it. It is headed up Dr. Martin Blaser, whose seminal work on the microbiome we discussed in chapter 5. But even this first-rate panel of experts could not come up with a workable recommendation for curtailing agricultural use. While noting that the Food and Drug Administration has made recent efforts to reduce antibiotic use in animals, requesting veterinary oversight and an end to using antibiotics to encourage growth, the members conceded that there was nothing mandatory about these efforts and there was little evidence that they had had any effect since they were introduced in 2012.

One of the panel members, Dr. Michael Apley, a veterinarian at Kansas State University and an expert in agricultural uses of antibiotics, advocates that all such use be left in the hands of veterinarians and calls for much more study of the issue. So far, we have essentially left these matters in the hands of vets, and made only limited progress.

Certain enlightened nations like Sweden, Denmark, and the Netherlands have limited agricultural use and set up comprehensive surveillance systems to determine the rates of antibiotic resistance in human and animal disease-causing germs. Dr. Jaap Wagenaar, professor of clinical infectiology at Utrecht University, points out that while the Netherlands has traditionally had the lowest rate of antibiotic use for humans in the European Union, as a major agricultural exporter, it was the highest on the animal side. To combat this, the health ministry set prospective standards to be met year by year, mandating full and transparent reporting by the industry. Antibiotics for animal use must be prescribed by licensed veterinarians. For the most powerful antimicrobial agents, there must be confirmation that there is no reasonable alternative to their use.

Most other nations have not attempted to institute such progressive practices. As the members of the developing world have adopted our meat-centric diet, they have also adopted our agribusiness formula for producing that meat, making heavy use of antibiotics for animal growth.

As a result, resistance is developing at an alarming rate. Fluoroquinolones (so named because of the fluorine atom in their central molecular structures) belong to a family of broad-spectrum antibiotics and include Cipro and other compounds whose scientific names end in "floxacin." In a 2016 presentation at NIH, Ramanan Laxminarayan, a widely respected economist and epidemiologist who specializes in research on the impact of infectious diseases and drug resistance, noted that in 1990, there was a 10 percent resistance rate in the common pathogens found in animal production. By 1996, the rate was over 80 percent.

For quite some time, many of us in the public health field have been attempting to determine just how widespread the use of antibiotics in animals is in the United States and what those antibiotics are used for, but the food-animal producers have been reluctant to give us figures or administration data. Large meat producers claim it is proprietary data and they are afraid it will be used to blame the industry for the rise of superbugs. Martin Blaser puts the annual use of antibiotics for animals at 14,000 tons, compared to 4,000 tons for humans. The mere fact that we have to use measures like total tons of antibiotics, which is such a crude estimate of use and doesn't tell us anything about types of antibiotics or where and how they are administered, is clear evidence that we desperately need better data. We believe antibiotic dosing for growth is being phased out in the United States, but how much is unclear. We do know that overall, according to various reliable sources, antibiotic use in American agribusiness is growing faster than livestock production. Between 2009 and 2014, antibiotic use increased by 22 percent.

I would liken our need for clear data on this to the need for hospitals in the United States to report the frequency of healthcare-associated infections in their institutions. Hospitals are now required by the federal government to report this data, but that wasn't always the case, and there was a great deal of reluctance and pushback by the hospitals when the requirement was proposed. Today, the reporting system is in place and is a major reason why hospitals are taking extra measures to prevent patients from becoming infected while being cared for in their hospitals. The details of antibiotic use in food animals, beyond the raw numbers, are vital public health information, and as far as I am concerned, that trumps proprietary claims any day. Without the information, we can't even establish a safe target for future use.

On May 10, 2016, the US Food and Drug Administration finalized a rule that revises annual reporting requirements for companies selling antibiotics for agricultural use. In addition to the overall estimates they now submit on the amount of antimicrobial drugs they sell to food-animal raisers, they must now break the number down by species: cattle, swine, chickens, and turkeys.

The FDA's statement promises, "The new sales data will improve the agency's understanding of how antimicrobials are sold and distributed for use in major food-producing species and help further target efforts to ensure judicious use of medically important antimicrobials."

This is all well and good and could help us get a handle on the agricultural dimension. But it took forty years to get even this far. We don't have forty more years for the rest of the world to get on board. Focusing only on reducing antibiotic consumption in the United States, Canada, and the European Union would be like patching three square feet of the twelve-foot-square hole the iceberg ripped in the *Titanic*'s hull and congratulating ourselves that we once again have a seaworthy vessel.

Use for Animals in the Rest of the World

Antibiotic use is growing rapidly beyond the First World and is already leading to huge problems. Blaser estimates that 81,000 tons of antibiotics per year are used in China for humans and an equal amount for agriculture. China also exports another 88,000 tons annually. In China and other Asian nations, serious regulatory oversight is virtually nonexistent. The New Delhi–based Centre for Science and Environment found that 40 percent of seventy samples of chicken meat bought in that city's markets from September 2013 to June 2014 contained antibiotic residue. Blaser has found no data he considers reliable for India.

We do have enough information to consider that India may be the largest producer of antibiotics in the world and, in turn, the greatest user and exporter of these drugs.

Maryn McKenna cites India and China as the largest

practitioners, with India "completely stuck in dysfunction on this." Many of her own findings were borne out by an investigation Bloomberg News undertook in 2016.

We see another frightening example of the mess we're in, in China, with the use of colistin, an absolute last-ditch antibiotic for bacteria that react to nothing else. It was isolated in Japan in 1949 and then developed in the 1950s, but it was not used unless absolutely necessary because of potential kidney damage. It's not being used for people in China, but it is being used in agriculture—thousands of tons a year. Likewise, in Vietnam it is approved only for animal use, but physicians obtain it from veterinarians for their human patients.

Colistin is used for people, though, in much of the rest of the world, including India. As other antibiotics with fewer harmful side effects have become resistant, colistin is about the only agent still effective against certain bloodstream infections in newborn infants. In early 2015, as reported by Bloomberg, physicians treating two babies with life-threatening bloodstream infections at King Edward Memorial Hospital in Pune, India, found that the bacteria were resistant to colistin. One of the babies died.

"If we lose colistin, we have nothing," stated Dr. Umesh Vaidya, head of the hospital's neonatal intensive care unit. "It's an extreme, extreme worry for us." Some hospitals in India are already finding that 10 to 15 percent of the bacterial strains they test are colistin resistant.

What is worse, some bacteria can share independent little hunks of DNA, called plasmids, with one another. On one such plasmid, Chinese researchers found a gene known as mcr-1 that conferred colistin resistance. More recently, they have detected NDM-1—for New Delhi metallo-beta-lactamase—an enzyme that protects bacteria against an important class of antibiotics called carbapenems, used mainly in hospitals against already multidrug-resistant bugs.

Dr. Jianzhong Shen, professor of veterinary medicine at the

China Agricultural University in Beijing, told Bloomberg reporters Natalie Obiko Pearson and Adi Narayan, "The selective pressure imposed by increasingly heavy use of colistin in agriculture in China could have led to the acquisition of mcr-1 by *E. coli*." This does not mean that all or even many of the countless *E. coli* strains around the world will take on resistance, but it is disturbing in its implications for how resistance is spreading through indiscriminate antibiotic use in agriculture.

Just as we were completing this book, the colistin-resistant *E. coli* made itself known in the United States—in the urine of a forty-nine-year-old woman in Pennsylvania. When an article documenting this unhappy development appeared shortly after in *Antimicrobial Agents and Chemotherapy,* a journal of the American Society for Microbiology, the CDC's Tom Frieden said, "It basically shows us that the end of the road isn't very far away for antibiotics—that we may be in a situation where we have patients in our intensive-care units or patients getting urinary tract infections for which we do not have antibiotics."

Many of the largest chicken-growing concerns in India, including ones that supply meat for the nation's McDonald's and KFC outlets, use one of several antibiotic cocktails that combine colistin with such other vital antibiotics as ciprofloxacin (Cipro), levofloxacin, neomycin, and doxycycline. According to an article by Pearson and Ganesh Nagarajan, "Interviews with farmers indicated that the drugs, permitted for veterinary use in India, were sometimes viewed as vitamins and feed supplements, and were used to stave off disease—a practice linked to the emergence of antibiotic-resistant bacteria."

"The combination of colistin and ciprofloxacin is just stupidity on a scale that defies all imagination," commented Dr. Timothy Walsh, professor of medical microbiology at Cardiff University in Wales.

In 2011, the Indian government released a document entitled "National Policy for Containment of Antimicrobial Resistance," which called for a ban on over-the-counter sales of antibiotics for humans and on nontherapeutic use for livestock. The recommendations caused such an outcry from industry stakeholders that they were quickly withdrawn.

What are the implications of all of this? The end result could very well be untreatable bacterial infections going directly into the world food supply. This would be the ultimate Frankenstein scenario.

CHAPTER 17

Fighting the Resistance

The odds of Ebola breaking out are quite low, but the stakes are very high. With antibiotic resistance, the odds are certain and the stakes are just as high. It is happening right under our noses.

—Joshua Lederberg, MD

Of the world's 7.3 billion people, the United States, Canada, and Europe represent about 869 million, or roughly 12 percent. You can throw in Australia and New Zealand, but that won't make much of a difference in the numbers. But we are important in other ways. Collectively, we dominate science. We dominate the development of new healthcare treatments and inventions. And we dominate the world market in the creation of new drugs, vaccines, and antimicrobials.

Once those pharmaceuticals go out of patent, their generic equivalents are largely produced overseas, more than half in India and China. They are then sold back to the United States, Canada, Europe, and the rest of the world. It is easy to see the interrelationship we all have in this area. And so it stands to reason that even though the United States and these other First World countries account for only 12 percent of the global population, the rest of the world will look to us before they commit to a policy and plan for dealing with antibiotic resistance. If we can't get it right for humans and animals in the United States, Canada, and Europe, how can we expect the rest of the world to follow?

My first publication on antimicrobial resistance was in the *New England Journal of Medicine* in 1984. It dealt with fatal drug-resistant salmonella infections. Since then, I have grown increasingly alarmed by the public health implications and challenges of drug-resistant diseases. I have studied the ever-worsening resistance issue for more than thirty years, and I have actively participated in professional organizations and government committees and work groups throughout that time, and I believe there are four priorities that must be addressed immediately to stem the growing antimicrobial resistance crisis in uses for both humans and animals. Some of them are expensive; some are virtually free. But all need to be implemented and none are pie-in-the-sky unrealistic. They are:

1. Preventing infections that require antibiotic treatment.
2. Protecting the efficacy of the antibiotics we currently have.
3. Discovering and developing new antibiotic agents.
4. Finding novel solutions that take some of the pressure off antibiotics.

Preventing Infections That Require Antibiotic Treatment

The first priority is where we've seen the most tangible progress, at least in the institutional environment. In 2013 the CDC outlined the top eighteen urgent, serious, and concerning antibiotic-resistance threats in the United States. Seven of the eighteen involve bacteria usually acquired in healthcare settings, including hospitals and long-term-care facilities. This should not be surprising, as more than half of hospitalized patients on any given day are receiving antibiotics and about one in twenty-five patients has one or more health-associated infections.

Controlling antibiotic-resistant infections associated with healthcare requires two separate actions: first, reduce antibiotic-resistance development by more judiciously using antibiotics; and second, prevent the transmission of antibiotic-resistant bacteria with improved infection control. We know how to be successful with both of these actions; there is not a great discovery waiting to be made. But getting the job done requires providing adequate resources and training, accurately measuring patient outcomes, and holding people accountable when cases of preventable resistant infections occur.

As we noted earlier, when hospitals were first required to report infection rates, many doctors and administrators threw up their hands and said, "This is going to ruin us!" It turns out to have been the greatest incentive we've had for infection control. Almost every hospital had an infection-control program prior to this, and some achieved laudable results. But improvements began to accelerate when the government either imposed financial penalties or provided incentives for performance-based accomplishments. Shrewdly, the Centers for Medicare and Medicaid Services started tying payments to patient outcomes. That one step has prevented the need to use a fair amount of antibiotics in the first place.

Other preventative measures are as simple as frequent handwashing. More than 160 years since Dr. Ignaz Semmelweis demonstrated to his Austrian medical colleagues that washing their hands before touching patients prevented hospital deaths, many medical personnel have yet to learn the lesson. According to most statistics, doctors are worse offenders than nurses.

On the international front, there has to be a major directed focus on providing clean water, basic hygiene, and sanitation—inadequate infrastructure elements such as these are huge promoters of infectious diseases—to places that don't have them. More than 2 million people die around the world each year from waterborne diarrheal disease. Contaminated water encourages the cycling of bacteria between humans and the environment and stimulates the dissemination of resistant genes.

If the infrastructure in each country were improved in terms of clean water and adequate sanitation, many of the antibiotic courses currently prescribed wouldn't be necessary.

A preliminary AMR report states, "Using data published by the World Bank and the World Health Organization, we have found that when income is controlled for, increasing access to sanitation in a country by 50 percent is correlated with around nine and a half years of additional life expectancy for its population."

In the same vein, the WHO suggests that universally giving pneumococcal vaccine to children under age five would save 800,000 yearly deaths from *Streptococcus pneumoniae*. A related study in the *Lancet* estimated that this step would also prevent the need for 11.4 million days of antibiotic use per year.

One truth I've observed throughout my career is that what gets counted is what gets acted upon. Therefore, I have always stressed disease surveillance: the science of finding and counting cases. This is critical. If we don't know about a disease or outbreak, we can't do anything about it. The CDC has a rapid-detection system for new flu strains and in July 2016 announced a $67 million program to begin the establishment of a similar system for antibiotic resistance in the United States.

About a year earlier, the World Health Assembly initiated GLASS: the Global Antimicrobial Resistance Surveillance System, to support a standardized approach to the collection, analysis, and sharing of data on a worldwide level. But this program is voluntary among member nations and there is no dedicated funding to support it.

In addition, there are three regional and partially overlapping networks—Latin America, Central Asia, and Eastern Europe; and Europe-wide—but funding is limited and so are the areas covered.

I consider all of these programs a down payment on what we ultimately need: a comprehensive, rapid-surveillance mechanism that could alert not only the United States, but all parts of the world when a new infectious disease emerges.

Such a surveillance system has the potential to halt a bacterial outbreak before it spreads. Not only would that prevent needless sickness; it could obviate the need for hundreds or thousands of antibiotic doses in each instance.

Protecting the Efficacy of the Antibiotics We Currently Have

If there is one word that weighs more heavily than any other in the discussion of preservation of our antibiotic arsenal, that word is not "science" or "research" or even "funding." That word is "*behavior.*"

From the perspective of medical standards and practices, the key to protecting the efficacy of our current antibiotics is what is known in our business as stewardship. Dr. Barry Eisenberg of Merck has characterized stewardship as "the right drug for the right patient at the right time for the right duration, with the right diagnosis." It means that there should be one expert infectious disease specialist or group at every hospital that controls the prescription of powerful antibiotics so they are not used inappropriately; if you want a particular antibiotic for your patient, you'd need permission from the infectious disease specialist.

Unfortunately, in many cases, this is easier said than done, as doctors seldom want to give up autonomy in patient care. From his perspective as a hospital clinician, Spellberg told us, "I've lost count of the number of people I've talked to who either run or are involved in stewardship programs at hospitals, who say, 'We'd love to do restriction programs, but we can't because the docs just won't tolerate it.'

"Then why are we asking them? The fundamental concept here is that if antibiotics are a societal trust—if my use affects your ability to use them, and then your use affects my grandkids' ability to use them—why are we allowing people to choose? We recognize in society that individual autonomy extends only up to the point that you begin to affect others."

Rarely, with stricter guidelines on the use of powerful antibiotics, we may make a fatal mistake. As the old sardonic punch line reminds us, medicine is not an exact science. Given the choice between "What harm am I doing to society in the future?" and "What harm might I be doing to my patient now?" Spellberg conceded that effective stewardship means that once in a while, a patient might die because he or she wasn't given an antibiotic, like the twenty-five-year-old woman with the fever, sore throat, and headache who died of septic shock a week after her hospital visit.

"I know that I'm going to do far more harm than good if I give 10,000 people inappropriate antibiotics to prevent one of those cases," he said. "But it's the ones you lose that psychologically stay with you, not the ones who did fine. And until we, as a society, grapple with that fear and the irrationality of our inability to proportionally assess risk, we're going to continue to abuse antibiotics."

Effective antibiotic stewardship must involve the public reporting of antibiotic usage by hospitals, medical services, and private practitioners, to embarrass and discredit those who overuse and misuse antibiotics. A recent study tracking antibiotic use among physicians whose prescription rates were published showed a significant downturn in antibiotic use. In private practice, this ultimately could lead to an adjustment of reimbursement rates from insurance carriers and the government.

Another strategy draws on a well-known psychological principle called "public commitment." Asking doctors to post statements in their exam rooms that say, essentially, "This office will not prescribe antibiotics to patients with viral infections because it is harmful and won't be effective," helps to ensure that both they and their patients understand and become comfortable with proper standards of care at the outset. Doctors don't want to go back on their stated word and patients come in with altered expectations. In offices and clinics where this has been tried, antibiotic prescriptions have gone down an average of 25 percent and patients feel as if they are a part of the overall effort to curb inappropriate antibiotic use.

As elementary as it may sound, three of the strongest psychological tools we have in the stewardship of prescribers to preserve our existing antibiotic weapons are public accounting and/or embarrassment, financial incentive and disincentive, and public commitment. If we use these tools widely and wisely, they will work.

For every pharmaceutical licensed in the United States, national guidelines are published for its use. Members of the Infectious Diseases Society of America (IDSA) and other experts are largely responsible for establishing these guidelines. Obviously, the drug companies want them to be as broad and inclusive as possible and to be able to market their products accordingly. And let's not kid ourselves—pharmaceutical marketing to physicians and hospitals is highly effective. Otherwise, the pharmaceutical firms wouldn't spend so much time, money, and effort on it.

So part of this effort on guidelines is to restrict the labeling on these antibiotics to prioritize their usage. You may ask, How big a deal is this? Do doctors actually read or respond to drug labels? No, the vast majority of the time they don't. But narrowing antibiotic usage guidelines on labels restricts what the pharmaceutical companies can market each one for. Unlike powerful psychotropic drugs used in psychiatry, where most of the inappropriate use is off-label, with powerful antibiotics, it turns out, most of the inappropriate prescribing is actually on-label use.

This is not as simple an issue at it might appear, although it should be. By statute, the FDA evaluates and licenses drugs based on clinical data that conclusively demonstrate safety and effectiveness. With antibiotics, this is clearly not enough. Congress needs to pass legislation to the effect that the FDA can restrict an antibiotic's authorized use to certain serious conditions so that labeling can be made to reflect that.

When national guidelines and/or product labeling say that a particular antibiotic that is one of the few effective treatments against really dangerous bacteria—say, *Pseudomonas* and *Acinetobacter*—can also be used against more common bacterial infections for which penicillin or erythromycin would suffice, doctors contribute to the problem.

Under existing circumstances, it is easy to see why the chief surgical resident in Spellberg's earlier story wanted to use the Zosyn. National guidelines told her she could. So let's add a significant and meaningful narrowing of these guidelines—a ranking of recommended antibiotics for each infectious condition—to our critical to-do list.

So far our recommendations apply mainly to the United States, Canada, and Europe. There is only a limited amount we can do to stop the rest of the world from squandering antibiotics. However, at the top of that list, it seems to me, is an international effort to convince foreign leaders, health establishments, and general populations that we're all in this together. What gives me hope in this regard is that it seems the international awareness and action effort on global climate change is starting to bear fruit. We need just such an education program worldwide for antibiotic preservation, just as we need a program similar in power to the decades-long antismoking campaign here in the United States.

Admittedly, as Maryn McKenna points out, this is not as simple or direct as the antismoking campaign, where we can say straight out that cigarettes are a devastating health enemy. We have to convey a far more nuanced message—that antibiotics are miraculous if used properly, but they shouldn't be used at all if they aren't really needed, and that even though we don't want to overuse them, we want patients to complete their prescriptions and not stop just because they feel better, and... Well, you get the idea.

The CDC has undertaken an antibiotic educational outreach of sorts, but for an issue as important to public health and as complex as this one is, McKenna suggests, we probably do need an effort as massive as the antismoking messaging, with government getting behind it.

The outreach effort on antibiotic stewardship in food animals will be more complex, largely because there is so much money at stake. But Ramanan Laxminarayan has studied the issue from both medical and economic perspectives, and he believes that as breeding technology progresses, antibiotics are playing a smaller and smaller role in animal growth. He says that if antibiotics were now withdrawn as a growth promoter from pigs in the United States, all positive and negative factors considered, the total economic impact would be a reduction of only $1.34 in the price per pig. If we can tackle this one issue—in pigs, cattle, and poultry—with the hard data to support it, we can begin to make a real difference.

We will continue to advocate for the *safe* and *appropriate* use of antibiotics for sick animals, those we raise for food and those we cherish for work, recreation, and companionship. But right now, we are a long way from that standard. Today, we are using antibiotics largely to clean up from and compensate for our unsanitary and overcrowded animal-production facilities. We need to correct these conditions for both scientific and humanitarian reasons. Experts like Laxminarayan are well equipped to figure out the economic implications.

I believe this is so vital that in 2016, we at CIDRAP launched a cutting-edge, web-based information platform for antimicrobial stewardship. The site provides the most current, comprehensive, and authoritative information on all aspects of the issue for the global community.

Discovering and Developing New Antibiotic Agents

Now we come to the issue of discovering and developing new and effective antibiotic agents. This is getting harder and harder as resistance grows, but it is not beyond our scientific capabilities. After all, in the three-quarters of a century that we've been at this, we've cultured only about 1 percent of the bacteria on the planet. We don't know how many more really good ones are out there waiting for us.

We can't expect the big, for-profit companies to handle the lion's share of new antibiotic development, because we can no longer rely on procuring our antibiotics from the traditional business model. Up-front costs and the time it takes to get through clinical trials and approval are key discouraging factors, as is opportunity cost. It is much more profitable for a big pharmaceutical concern to devote its financial and development resources to a drug that people are going to take every day than to one that is used only rarely and that will be rationed in order to preserve its effectiveness.

In July 2016, BARDA, the Wellcome Trust, the United Kingdom's AMR Centre of Alderley Park, and Boston University School of Law announced the creation of "one of the world's largest public-private partnerships focused on preclinical discovery and development of new antimicrobial products." BARDA is providing $30 million for the project's first year and the AMR

Centre will contribute $14 million in the first year and up to $100 million over five years. Additional organizations will participate. The objective of this partnership is to "identify promising candidates in the early stages of development that may offer treatment options for drug-resistant bacterial infections."

This is certainly a promising start, but it is only a start. It seems like a lot of money is being devoted to this effort, but let's put it in perspective. A number of highly respected experts have called for an international scientific effort similar to that expended on CERN, the European Organization for Nuclear Research, which operates the largest particle physics laboratory in the world, with the aim of probing the fundamental structure of the universe. In an article in the January 12, 2016, issue of the *Lancet Infectious Diseases,* twenty-four distinguished scientists, led by Dr. Lloyd Czaplewski, pointed out that CERN's Large Hadron Collider project cost about $9 billion and the International Space Station cost about $144 billion, then concluded, "Antimicrobial research and development to address the problem of antibiotic resistance probably needs an investment that is somewhere between the two."

That's unlikely to happen, but it gives us some idea of the magnitude leading experts attach to the problem, though AMR's estimate of 300 million deaths and a $100 trillion loss to the world's economy by 2050 should get everyone's attention on its own.

What we suggest (as we did for vaccines) is the defense contractor model, and if antibiotics are a national trust, then this certainly makes sense. This model also puts some of the decision making into the hands of the public's representatives, as is the case in the defense industry. If the Pentagon decides it needs a new aircraft carrier, fighter jet, or any other class of equipment, it asks for bids and then awards a development contract.

In the case of the fighter jet or aircraft carrier, the government is going to be the only buyer. This will not be so with new antibiotics, though through Medicare, the military, the VA, and other programs, the government is likely to be a major buyer. The key thing the private-public partnership in antibiotic development does is take those major financial and present-value time pressures off the contracting pharmaceutical companies. In return for restricted-use labeling, the company can charge a premium price for the antibiotic in those situations where it really is the agent of first choice.

While we all complain about the cost of certain prescription drugs, in a case like this, we have to factor in the concept of true value. If a new antibiotic that costs considerably more than its generic predecessor can get a patient out of the hospital two or three days earlier, its true value must be weighed against the cost of those extra two or three days. Likewise, if the reason for the high cost is that the new agent is being held back from general usage so it won't lose its effectiveness against some otherwise untreatable bugs, the true value is almost beyond cost evaluation.

Still, Maryn McKenna adds a prescient warning: Even if we follow this model, "at some point, someone is going to come up with some financial mechanism that allows new drugs to flow into the marketplace. And if we don't change our behavior, we are going to use those drugs up as soon as we used up the old ones. Unless we change our behavior, we are never going to get out in front of the problem."

Finding Novel Solutions That Take Some of the Pressure off Antibiotics

How do we find novel solutions to the resistance problem? By figuring out how to prevent and treat some infections in ways that do not promote resistance.

First and foremost we need to prioritize basic vaccine research and development that address current or emerging antibiotic infections.

Also promising are host-modifying therapies. That means, rather than trying to kill the bug, the treatment would involve doing something with the host—the patient's body—that retards the infection. In some cases, this might mean blunting an inflammatory response. In others, it might mean enhancing it.

Another approach is treating some infections passively. For those bacteria that do their damage by releasing a toxin, such as staphylococcus or diphtheria, if you can neutralize the toxin, that's as good as killing the pathogen. One form of this method actually hearkens back to preantibiotic days: Serum therapy, invented by German doctor Emil von Behring in the 1890s as a treatment for diphtheria, involves injecting blood serum from someone who has already had the same infection into the patient.

Another passive strategy is to deprive the offending bacteria of nutrients it needs to divide and grow, such as iron. The bacterium cannot manufacture iron, so it must steal it from the host. If we can find ways to "hide" our iron from them, we might not have to attack the biochemical pathways in the bacteria, which is what allows the bugs to build up resistance. This is one area in which we might expect significant scientific breakthroughs in the coming decades.

Then there is the use of bacteriophages, which are viruses that can infect and kill certain bacteria. Lysins are enzymes produced by phages that digest the bacteria's cell walls. In other words, we treat the patient by intentionally introducing a virus that infects only the disease-causing bacteria. This concept has been understood for quite some time, but it has never really been tested, as it should be, in rigorous clinical trials. Again, this is a situation in which we need more and better data.

The AMR report also predicts that major advances in computer science and artificial intelligence could both crunch a lot of big data to determine the shortest effective time period for taking antibiotics for a given condition, and also help doctors with initial diagnoses. Applications could be directed toward analyzing agricultural use as well.

Finally, the development and implementation of rapid diagnostic and biomarker tests could help distinguish between viral and bacterial infections, whose resemblance, as we've seen, is at the basis of so much of the cautionary overprescribing. Such testing could also be extremely useful in disease surveillance. Many experts agree that the technology exists for this, but the financial incentives to develop and produce it may not. It all depends on what Medicare and the insurance companies are willing to pay for. For example, if the test costs more than the antibiotic that would be prescribed in the event that the test yielded a positive result, there could be significant pushback against it. On the other hand, if we are at the point where we've used up many of our cheaper agents, the rapid test becomes much more economical, even if its price hasn't changed at all.

We are starting to see some more awareness regarding the antimicrobial resistance threat on the international front. In April 2016, health ministers from twelve Asian-Pacific nations met in Manila under the auspices of the World Health Organization, the government of Japan, the Food and Agriculture Organization of the United Nations, and the World Organization for Animal Health.

After a two-day meeting they pledged mutual collaboration in combating resistance, conceding, according to a statement by WHO Western Pacific Regional director Dr. Shin Young-soo, "Antibiotic resistance is one of the biggest threats to human health today. Having effective antimicrobials is also critical to the social and economic development of nations. We have a limited window of opportunity to take action and avoid a post-antibiotic era."

If there is any serious prospect for dealing with the antimicrobial resistance issue in a comprehensive and international way, it can be found in AMR's May 2016 report *Tackling Drug-Resistant Infections Globally: Final Report and Recommendations.* There are no great surprises, but we can only hope that the credentials and reputation of the authors and the organization itself will put the necessary impetus behind its message.

The AMR report drills down on each of our four priorities, including raising awareness globally, improving sanitation and water quality, regulating agricultural antibiotic use, heightening surveillance, investing in rapid diagnostics, looking for alternate therapies, supporting treatments that aren't commercially viable, encouraging investment in new antimicrobials, and forming a global coalition for antibiotic stewardship.

More than half of the recommendations apply equally to all other important aspects of world public health, so it is not a question of devoting large resources to warding off a crisis that might not arrive. These initiatives will not only help us maintain the effectiveness of antimicrobial agents; they will help improve world health in general. What could possibly be more important than that?

The AMR authors recommend development of successive ten-year targets to reduce antibiotic use in farm animals, increase focus on food animal–raising practices, cease the use of last-line antibiotics that treat critical infections in humans, and require that food producers provide information about their antibiotic use, not only to the government, but to the public as well. If food sellers have to label whether their meats, poultry, and fish are raised with antibiotics, food buyers will certainly register their preferences in the retail marketplace, particularly if their choices have been supported by an awareness campaign.

The AMR report estimates that all ten programs will cost $40 billion over the next decade, but that the cost would represent a small fraction of the approximately $100 trillion in global production that would be lost due to drug-resistant infections predicted to occur by 2050.

The authors concede "that no single country can solve the AMR problem on its own and several of our proposed solutions will require at least a critical mass of countries behind them if they are to make a difference." For instance, if either China or India fails to participate or ante up, a lot of these proposed solutions are not going to work.

Not an easy task, probably no easier than it has been to galvanize the world regarding climate change. We can argue about how likely it is that these provisions will be accepted and acted upon. What is beyond argument at this point is what will happen if we do nothing, or not enough.

Jim O'Neill is cautiously optimistic that the commission's recommendations can succeed. His first encouragement, he says, came at the 2015 G20 summit in Antalya, Turkey, where a commitment to dealing with antibiotic resistance was included in the closing statement. "From my experience in finance," he says, "when something gets on either a G7 or G20 agenda, it's pretty rare that it disappears until they've actually done something about it. There are now a number of moving parts wanting to play more of a role at the same time.

"My dream would be a statement saying, 'The G20 ministers agree today that they will now work to implement the details of what they concluded in supporting a market entry reward system for new drugs, and the establishment of a new global fund to pay for these rewards.' "

O'Neill is also heartened by the statement put out by the pharmaceutical industry at the World Economic Forum in Davos, Switzerland, in January 2016. There, more than eighty leading international pharmaceutical, generics, diagnostics, and biotechnology companies, as well as key industry bodies, came together to call on governments and industry to take comprehensive action against drug-resistant infections—so-called superbugs. Whether the Davos statement is mere corporate lip service or will actually move the needle remains to be seen.

This commission and its recommendations represent our best shot. If we fail to grasp hold of this opportunity, we should be prepared to explain to our grandchildren why they have to live and learn how to survive without the protection of antibiotics.

CHAPTER 18

Influenza: The King of Infectious Diseases

Of all the things that could kill more than 10 million people around the world, the most likely is an epidemic stemming from either natural causes or bioterrorism.

— Bill Gates, *New England Journal of Medicine,* April 15, 2015

The public doesn't get worked up about the seasonal virus infection commonly known as flu the way we have for, say, Ebola and Zika. Yet the influenza virus causes a wide spectrum of conditions and consequences ranging from infection without any symptoms all the way up to death. In fact, in any given year, seasonal flu claims 3,000 to 49,000 lives just in the United States. That means in some years it causes as many as or more deaths than automobile accidents. Admittedly, many are among the elderly, the immunocompromised, or those in poor health to begin with. But as we do with highway fatalities, we seem to have factored the yearly influenza death toll into our individual threat matrices and decided there is little we have to worry about. Many of us don't even bother getting flu shots, even when they are offered at low cost at our local drugstores and may offer moderate protection against the illness in some years.

The reason we need a new vaccine formulation each year is

because influenza viruses that are transmitted between humans are unstable and unreliable. They mutate easily as they pass from person to person.

Influenza viruses, which belong to a family that has a single segmented RNA genome, are divided into different types, A, B, and C, based on their core proteins. As is characteristic of many RNA-genome viruses, these undergo high mutation rates and frequent genetic reassortment as they reproduce. Mutation occurs when the virus makes a "mistake" while reproducing itself in a single lung cell. Reassortment occurs when two different influenza viruses infect a human or pig at the same time and subsequently swap and rearrange genetic material to create a new hybrid virus.

Mutation of influenza viruses usually results in minor changes in the emerging strain that nonetheless require the vaccines to be updated, sometimes annually. When we describe virus mutation, we call this antigenic drift, a relatively small change. With reassortment, major changes occur, resulting in a new virus that can be unlike anything that humans have experienced before and can become the viral strain that starts the next worldwide pandemic. This process is referred to as antigenic shift. And because of all of this genetic shift and drift, the immune system often will have to deal with each new strain as something it hasn't seen before and so must mount a new attack.

We classify the type A influenza strains—the ones that cause influenza pandemics in both animals and humans—by the characteristics of two proteins on the virion's surface: hemagglutinin (HA) and neuraminidase (NA). The hemagglutinin has the ability to bind with lung cells it comes in contact with, like a key fitting into a lock, and that is what starts the viral reproduction process. When the cell's genetic machinery has churned out so many influenza virions that it's full to bursting, it does burst, and the thousands of new virions move out to bind with other cells. The purpose of the neuraminidase is to allow those virions to escape the cell's confines and spread to other cells, and even get expelled in the "wind of a cough." The antiviral drugs that work against most influenza strains—oseltamivir (brand name Tamiflu) and zanamivir (Relenza)—work by obstructing the function of the NA, which is why they are called neuraminidase inhibitors.

When we describe type A influenza viruses as H3N2, H1N1, or H5N2, we are referring to their HA and NA components. Technically we refer to influenza viruses by their type and HA and NA characteristics, such as A(H3N2). But for the type A viruses, the ones that cause influenza in humans and animals, we just shorten the name to the HA and NA components, for instance, H3N2. At present, we have identified eighteen distinct type A HA subtypes and eleven NAs, for a total of 198 possible combinations. The most recent pandemic, in 2009, was classified as H1N1—a descendant of the deadly 1918 strain.

Just as there are at least seventy-four different Donald Petersons in the Minneapolis white pages, two different influenza viruses with the same HA and NA may actually be different strains. For example in 2009, there was an H1N1 virus circulating in humans, as its forefather had been doing since 1977. But then a new and different H1N1 virus emerged in Mexico, most likely coming from a reassortment event in the swine population. Previous infection with the older H1N1 strain did not protect humans against the new strain, which resulted in the 2009–10 human influenza pandemic.

"The first thing to understand about influenza," says John Barry, author of the definitive account of the 1918 pandemic, *The Great Influenza,* "is that it's all bird flu; there's no such thing as a naturally occurring human influenza virus." The primary reservoir—meaning source—for type A influenza is wild aquatic birds. Birds can, and do, travel all over, so that spreading the virus, both respiratorily and through their droppings, is easy. Animal influenza viruses do not spread easily into humans. But they *can* readily spread to other species, including domestic birds such as chickens and turkeys, as well as dogs, cats, horses, and pigs. Pigs are especially important to infecting humans with avian influenza viruses. The cells lining their lungs have receptors that match up with both bird and human viruses, so those lungs turn out to be perfect places for influenza strains to "meet each other" and mix. It is even possible to have a triple reassortment, where strains of all three species—humans, birds, and pigs—mix to form a completely unpredictable new influenza virus. When that happens, it's a spin of the genetic roulette wheel whether the new strain is more or less serious than the strains from which it emerged. In 1918, that spin resulted in the virulence jackpot.

As far as pandemic potential is concerned, the most dangerous places on earth are anywhere people, birds, and swine are crowded close together in large numbers—the food markets of China and Southeast Asia, for example, or the industrial farms of the American Midwest.

It is the range of possible results from the changeability and mixing of influenza strains that makes it the king of infectious microbial beasts. While it can be almost as mild as a common cold, it can also be just as fearsome and deadly as smallpox, and even easier to catch. That is why this particular beast terrifies epidemiologists.

There's another crucial difference between influenza and all the other "maybe" point-source diseases, such as Ebola or Marburg, that form the basis of every plague novel and outbreak movie. As infectious disease epidemiologists, we all know that pandemic influenza is the one infectious disease that *will happen*.

It *has happened* at least thirty times since the sixteenth century, and our modern world presents all the ingredients for an imminent return.

As we mentioned earlier, no disease outbreak of modern times compares to the 1918–19 worldwide influenza pandemic.

Though it was called the Spanish flu, it may have started in the United States, specifically in Haskell County, Kansas, in an agricultural setting. Whether this particular strain began in pigs and spread to humans or vice versa is not clear. Epidemiological evidence suggests that from Kansas it probably traveled east to the large army base at what is now Fort Riley, and then went with the recruits to Europe. The high concentration of soldiers living in close confines as they trained for combat in the Great War certainly exacerbated the situation, as did the large-scale movement of troops across the oceans.

Unlike most seasonal influenza virus strains, the 1918 H1N1 strain was anti-Darwinian: Rather than claiming the old, the infirm, and very young children—those with weak or underdeveloped immune systems—this one killed off the strongest and fittest, as well as pregnant women, in disproportionately high numbers, causing a "cytokine storm" in healthy individuals, as we described in chapter 5. This immune system overreaction critically damages the lungs, kidneys, heart, and other organs. Today, we are not much better at treating patients dying from a cytokine storm than we were in 1918. The 2009 H1N1 pandemic did not cause a large number of human deaths, but a number of those it did kill were younger adults in whom the flu triggered a cytokine storm, just like in 1918.

In 1918–19, those deaths were grisly. Within hours of the victim's first onset of symptoms, blood would begin to leak into the air spaces in the lungs. By the second day, the lungs had been transformed from an oxygen-rich "sponge" to a bloody "rag," the suffering patient literally drowning in his or her own fluids. "One robust person showed the first symptom at 4:00 pm and died by 10:00 am," notes a contemporary report.

Those who didn't succumb to the cytokine storm were still susceptible to deadly or fatal pneumonia caused by a secondary infection; bacteria were able to infect the lungs because the initial flu virus had destroyed the protective epithelial cells that lined the breathing passages. We can't retrospectively separate out the viral deaths from the subsequent bacterial deaths, but the indications are that most of the morbidity and mortality was from the initial virus, so even if they had had antibiotics in those days, they wouldn't have been of much use.

In New York City, the pandemic left 21,000 children orphans. It was so widespread that the disease peaked in Boston and Bombay at the same time. In some parts of the world, according to John Barry, the death rate was so overwhelming that it was impossible to bury all the corpses. At one time or another, almost every city in the United States ran out of coffins. Ordinary civic and commercial functions were not being carried out because so much of the workforce was sick or dead. Some sick people starved to death, not because there was a food shortage but because so many people were afraid to come in contact with them. Unlike a virus such as Ebola, which is not communicable until the victim starts having symptoms, with influenza, you're contagious before you even feel sick.

The latest estimates suggest that the worldwide death toll may have reached 100 million—far more than all of the soldiers and civilians killed in the First World War. The bubonic and pneumonic plagues of fourteenth-century Europe took out a larger proportion of the smaller population of the time, but in sheer numbers of human beings killed, the 1918 flu was the deadliest single pandemic killer of all time. More people died in a six-month period over the fall, winter, and spring of 1918–19 than have died from AIDS in the roughly thirty-five years since that virus was identified in the human population.

So profound were the effects of the outbreak that the statistical average life expectancy in the United States was immediately lowered by more than ten years. Keep in mind that the world's population in 1918 was about a third of what it is today.

Amid the annual seasonal flus that have crept up each year since then, there have been three influenza pandemics: 1957

H2N2 Asian flu; 1968 H3N2 Hong Kong flu; and 2009 H1N1 swine flu. None of these came close to causing the devastation of the 1918 influenza, but the worldwide morbidity and mortality were still significant. In 2009, public health officials actually had been on the lookout for a spread of H5N1, a strain out of Southeast Asia that had so far not transmitted from person to person, but when it went from animal to human, the mortality rate was as high as 60 percent.

Back in 1976, after several soldiers fell ill and one died at Fort Dix, New Jersey, from what appeared to be an H1N1 influenza strain that closely resembled the 1918 strain, public health officials decided to take no chances and urged President Gerald Ford to authorize a mass, publicly funded vaccination program. At that time, there were a large number of people alive who had personally experienced the 1918 pandemic. It turned out that the 1976 epidemic did not materialize and there was no disease beyond Fort Dix. The aftermath of the vaccination campaign and the associated Guillain-Barré syndrome cases left a legacy of mistrust and skepticism that, to some extent, we are still fighting today.

In retrospect, it is difficult to fault the public health officials who were so alarmed when they saw evidence of H1N1 in the soldiers at Fort Dix. But if we had it to do all over again—and at some point in time we will—what we *should have done* was ramp up the vaccine production and then wait to see if the virus started to spread before undertaking a massive inoculation effort.

When the H1N1 virus from the 2009 pandemic was analyzed by Dr. Robert Webster and his colleagues at St. Jude's Children's Research Hospital in Memphis, Tennessee, it was found to be derived from a North American swine influenza virus that acquired two gene segments from the European swine lineages.

As it turned out, the 2009 pandemic was considered relatively mild by most, though that was not the situation for many.

Globally, it is estimated that 300,000 people died from H1N1 infection, 80 percent under the age of sixty-five. The CDC determined that in the United States more than 60 million cases of infection occurred in the first year of the H1N1 pandemic and 12,000 people died. Of note, 87 percent of the deaths in the United States were in those under sixty-five years of age. This is in sharp contrast to the greater than 90 percent of deaths in those sixty-five and older that occur in a typical seasonal influenza year. So while the number of deaths was comparable to the number in an average flu year, the average age of those who died was much lower. The "victims of preference" in 2009 were pregnant women, obese individuals, those with asthma, and those with certain neuromuscular diseases; they accounted for about 60 percent of the severe or fatal cases. This pattern of deaths is very similar to what the world experienced in 1918, just on a much smaller scale.

We now realize that there are two distinctly different patterns of influenza pandemic cases. One is what we saw in the 1918 and 2009 pandemics, where severe illness and deaths fall disproportionately on young adults. The second is what we saw in the 1957 H2N2 and 1968 H3N2 pandemics, where most of the deaths were in the older population, as is the case with seasonal flu. The average ages of death in the 1918 and 2009 pandemics in the United States were 27.2 years and 37.4 years, respectively. When considering that life expectancy in 1918 was forty-eight years and in 2009 it was seventy-eight years, the deaths in 2009 actually reflected an even younger demographic than did those in 1918. In the 1957 and 1968 pandemics, the average ages of death were 64.2 and 62.2 years, respectively. These ages are close to life expectancy at the time; in 1957 life expectancy in the United States was sixty-eight years and in 1968 it was seventy.

When our research group calculated a measure of early death for the three twentieth- and one twenty-first-century pandemics—a statistic known as "years of life lost before age sixty-five," we found that the 2009 pandemic had a much greater human impact than is reflected by total number of deaths alone. This is an important consideration when planning for future pandemics, as the impact on our healthcare resources and the workforce of the global economy will differ dramatically between pandemics where most of the serious illnesses and deaths are in younger adults, and those that primarily affect the older, largely retired population. Unfortunately, the average age of death for current cases of H5N1 and H7N9, two of the leading avian influenza virus candidates for the next pandemic, is in the early fifties.

Even a moderately severe pandemic would impact just about every aspect of our lives.

We have a global just-in-time-delivery business model and everything we use today is connected in some critical aspect to a production line far distant from our homes. If a factory in China suddenly can't function because 30 or 40 percent of its workforce is sick, we don't have a stockpile of its goods waiting in a closet or warehouse to tide us over until the factory reopens. If we have a similar outbreak in enough places at the same time, and factories can't get the parts and supplies they need from other factories, then we start to see a domino effect in which world trade suffers and economies start to falter.

And it is not just trade. If that same percentage of workers is off the job for days or weeks, then cities start to have trouble functioning. The trash doesn't get picked up, there aren't enough firefighters to fill each shift, police officers can't respond to every call, schools close down, and doctors and nurses don't show up at hospitals.

Hospitals and healthcare systems will suffer most acutely. As long as the number of cases doesn't exceed the capacity of our intensive care units, these units will be able to help patients who present with severe influenza symptoms. But what if the number of severe cases goes up by 30 percent? Guess what: We're pretty much at capacity now under normal circumstances, having cut all of the "fat" out of the system for budgetary reasons. We don't have any surge capacity. We also will run out of the equipment we need to protect healthcare workers, such as respirators and the tight, face-fitting masks. Who will come to work if they realize they are substantially increasing their chances of catching influenza because of a lack of protective gear?

Here's an even grimmer example. If 1 percent of those critical influenza victims need ventilators, we can probably handle it. If 3 percent need them, forget it; we just don't have enough machines in the country, and neither does any other country. Even if they did, do you think they would lend them to us? That means a lot of people would die even though we have the technology to save them. We'd get into triage and issues of allocation and hard choices no one wants to confront.

Shortly before the 2009 outbreak, we conducted a study at CIDRAP in which we surveyed a world-class group of pharmacists who had expertise in the drugs used in the various hospital medical specialties, such as acute care, chronic care, emergency care, and so on. We asked them what drugs they absolutely had to have on a day-to-day basis. Not cancer drugs, not AIDS drugs, but the essential, needed-to-sustain-life-can't-wait-until-tomorrow drugs. We ultimately compiled a list of more than thirty such critical pharmacological agents, including insulin for type 1 diabetics; the vasodilator nitroglycerine; heparin for blood thinning and dialysis; succinylcholine for muscle relaxation during surgery, intubation, and heart-lung machine hookup; Lasix for congestive heart failure; metaprolol for angina and severe hypertension; norepinephrine for severe hypotension; albuterol to open airways in the lungs; and various other heart and blood circulatory drugs and basic antibiotics.

One hundred percent of these drugs were generic; all were manufactured primarily or exclusively overseas, mostly in India and China; there were no significant stockpiles, and the supply chains were long and extremely vulnerable.

We must not think of the potential human pain and suffering from an influenza pandemic as limited to those who develop infection here in the United States. We must realize and plan for the terrible impact a pandemic could have and all the deaths that would occur as a result of an acute shortage of lifesaving drugs or medical care. And it should matter greatly to us if a factory worker in China or India who is responsible for helping to manufacture these drugs is too sick to work or a freighter ship captain who is delivering them dies en route.

Today, influenza is hyperevolving, more so than at any other time in the earth's history. The huge number of animals needed to produce our food serves as the amplifying factor for virus transmission and, in turn, for more spins at the genetic roulette table. Recall in chapter 17 on antimicrobial resistance, we described the need to feed 7.3 billion people in today's world. The rapid recent expansion of modern confinement agriculture, together with the establishment of many millions of smaller farms around the world, has given influenza viruses every opportunity to find suitable hosts to proliferate in poultry and pigs. The 88,723,000 metric tons of annual global poultry meat production equates to many billions of birds hatched, raised, and slaughtered. All of these birds have frequent direct or indirect contact with humans. In addition, the 413,975,000 swine produced globally add the last—and perhaps the perfect physiological—ingredient to the influenza virus evolution process.

In February 2015, the WHO issued a document entitled "Warning Signals from the Volatile World of Influenza Viruses." The report warned about the rapid changes in potential human pandemic strains in birds:

> The diversity and geographical distribution of influenza viruses currently circulating in wild and domestic birds are unprecedented since the advent of modern tools for virus detection and characterization. The world needs to be concerned.
>
> Viruses of the H5 and H7 subtypes are of greatest concern, as they can rapidly mutate from a form that causes mild symptoms in birds to one that causes severe illness and death in poultry populations, resulting in devastating outbreaks and enormous losses to the poultry industry and to the liveli hoods of farmers.
>
> Since the start of 2014, the Organisation for Animal Health, or OIE, has been notified of 41 H5 and H7 outbreaks in birds involving 7 different viruses in 20 countries in Africa, the Americas, Asia, Australia, Europe, and the Middle East. Several are novel viruses that have emerged and spread in wild birds or poultry only in the past few years.

This statement summarized thirteen months of increased virus activity from January 2014 to February 2015. Just thirteen months later—March 2016—the number had grown to hundreds of H5 and H7 outbreaks, involving nine different viruses in thirty-nine countries.

This frightening growth in H5 and H7 activity does not necessarily mean a human pandemic is imminent. But it could be. Of the 850 reported sporadic cases of human H5N1 infection documented since 2004, 445, or 52 percent, have resulted in death. The average age of those infected has been early fifties, substantially lower than the average age seen in seasonal influenza deaths.

For H7N9, 212 individuals, or 37 percent of reported cases, have died since this infectious strain was first documented in 2013. The average age of these cases has been around fifty. And there are more H and N type A avian influenza strains of concern in addition to H5N1 and H7N9. H5N6 has been circulating since 2013 in poultry in southern and western China, Laos, and Vietnam and has caused recent cases in humans. The list of these avian influenza viruses with the potential to infect humans continues to grow.

In 2015, high-pathogenicity (causing severe and fatal disease) avian H5N2 came to our own backyard here in Minnesota, as well as to other parts of the central United States. From early March to mid-June an unprecedented outbreak of an H5N2 strain occurred in Upper Midwest poultry farms. Two hundred twenty-three farm operations were infected; more than 48 million birds died or were euthanized. This virus likely arrived in the Midwest with migratory birds from Asia, possibly via birds sharing viral strains in the Mississippi and Rocky Mountain flyways.

It remains unclear how the H5N2 virus moved so quickly among facilities miles apart. I was the senior investigator of a large epidemiologic study trying to understand how the virus spread from farm to farm. Despite our efforts, we are still unsure what happened. Personally, I believe that after the virus-infecting wild birds came into contact with domestic poultry, the virus spread via humans through contaminated clothing and boots on people moving between facilities, or through sharing contaminated equipment; or via airborne transmission as the poultry shed substantial virus before they died and that virus-contaminated air escaped outside of the barns.

The H5N2 outbreak was a disaster for the poultry industry and could have been a first step toward a new human pandemic. Many of the same counties where the poultry outbreaks occurred have some of the highest populations of confined swine operations in the Midwest. Remember, when pigs get infected with influenza viruses, they rarely show many symptoms. But they can become infected with both avian influenza viruses and human influenza viruses simultaneously, and their lungs provide an ideal mixing bowl. With likely airborne transmission of H5N2 of up to miles from the source, and the colocation of the pig and poultry operations, I'm convinced the pigs were getting infected, too. They just didn't get sick or get tested for influenza infection. But in terms of what could happen, I'm convinced it's just a matter of time.

I believe I know less about influenza now than I thought I did fifteen years ago, even though I have been studying it continually since then. The more we learn about this virus, how it interacts with animal and human populations, how and why it changes genetically, and what those changes mean, the more questions we face and the fewer answers we can be sure of.

As a result, we can never be sure how close we are to the mutation or evolutionary pressure that will lead us to the next pandemic.

CHAPTER 19

Pandemic: From Unspeakable to Inevitable

> *And now was acknowledged the presence of the Red Death. He had come like a thief in the night. And one by one dropped the revelers in the blood-bedewed halls of their revel, and died each in the despairing posture of his fall. And the life of the ebony clock went out with that of the last of the gay. And the flames of the tripods expired. And Darkness and Decay and the Red Death held illimitable dominion over all.*
>
> — EDGAR ALLAN POE, *The Masque of the Red Death*

When we attempt to assess the risk of another 1918-type influenza pandemic, keep in mind the points we made earlier: that we live in a globally interdependent world, with widespread rapid travel and many concentrations of people, pigs, and birds living in close proximity. Thus, that world has become a hyper-mixing vessel—one with about three times the human population of 1918.

We don't know which, of all the influenza strains we're watching, will emerge as a pandemic one, or whether it will be something we've never seen before. What we do know is that when it happens, it will spread before we realize what is happening. And unless we are prepared, it would be like trying to contain the wind.

Larry Summers, a world-renowned macroeconomist in addition to being a former secretary of the treasury, provides a poignant perspective on this very point in his keynote address accompanying the release of the National Academy of Medicine's Global Health Risk Framework Commission report, *The Neglected Dimension of Global Security—A Framework for Countering Infectious-Disease Crises:*

> Of all the issues before us, pandemic and epidemic is the issue with the highest ratio of global seriousness to policy attention: that relative to its significance for humanity, there is no issue that gets less attention. To put the comparison in a direct way; that if you calculate the expected cost to humanity over the next century from epidemics and pandemics, on our current global path, it is in the same broad range, within a factor of two or three, as the expected cost from global climate change. And I am struck by how little attention this issue receives relative to the issue of global climate change.
>
> To be absolutely clear, global climate change deserves all the attention that it receives, and more. But I believe that global health risks deserve much more attention than they are receiving.

Our civil defense structure is set up for one-hit disasters, like an F4 tornado in Kansas, a Category 5 hurricane in New Orleans, or even airplanes hitting skyscrapers in New York. But what if we had twenty or thirty 9/11s or Hurricane Katrinas all at once? We wouldn't have the resources to handle that. As Defense Secretary Donald Rumsfeld so notoriously said of fighting the war in Iraq, "You go to war with the army you have. They're not the army you might want or wish to have at a later time."

A catastrophic influenza pandemic will unfold like a slow-motion tsunami, lasting six to eighteen months.

In 1918, there were three distinct waves of disease over a

two-year period, and that is what we could face again. So the only Hail Mary we would have is whatever we put in place beforehand.

Over the years, our team at CIDRAP has developed and led many "tabletop exercises" for organizations ranging from the White House and Fortune 500 companies to state and local governments, including public health departments and hospitals. These exercises are essentially simulated, realistic drills of disaster scenarios involving leaders in all areas of emergency management, public health, and emergency response to stress-test the plans a municipality, state, national government, or any other organized system has in place.

What follows is a fictional tabletop-like scenario involving an influenza pandemic in today's world with the virulence of 1918's H1N1 strain. It is narrated mainly in the present tense, as I would do when leading a tabletop exercise, with diversion into the past tense when information or historical perspective is needed. This scenario has been reviewed by colleagues in public health preparedness and business continuity planning. There is general agreement that it is realistic and possible. Keep that in mind as you imagine yourself and your family living through it.

At first, the doctors in the Shanghai metropolitan area think they're just seeing late-season flu cases, but their patients don't seem to be getting better. It is mid-April; influenza should be on the wane in China. It doesn't take long for physicians to realize that the hundreds of patients they are seeing in emergency rooms are presenting with conditions very different from anything they have seen before. At least fifty patients have died of acute respiratory distress syndrome (ARDS) in the past two days; intensive care units in many hospitals in the area can no longer admit new patients—they are bursting at the seams. In many cases, the victims report they had been sick for only a day or two, sometimes only hours. The majority of the victims are otherwise healthy young adults and pregnant women.

Clinicians quickly recognize that these patients have a similar devastating illness to the 1,000-plus Chinese who had been diagnosed with one of the bird flu infections in the past several years. Still, this is different: In the past, bird flu cases occurred only sporadically by location and time, rarely with multiple cases in one family. Now, emergency rooms and even intensive care units in hospitals all over the Shanghai area are awash in desperately ill patients.

The worst fears of Chinese public health officials are realized when sputum samples from eight patients hospitalized in three different facilities are confirmed to have H7N9 influenza infection. H7N9—an avian virus by origin that made its first recognized foray into the human population of China in 2013—has now taken the last major step to becoming the pandemic influenza virus.

Meanwhile, more cases are popping up in other places. In areas of China where this strain has been previously detected, about a third of those who contracted the disease from poultry have died. But the birds carrying the virus don't get sick or, at least, they don't display any noticeable symptoms. Within days, cases of H7N9 influenza begin showing up in hospitals throughout much of China and even other countries in Asia. Many of the first cases outside of Shanghai had recently traveled to the city. This story has gone from one of relative obscurity to the number one news story in the world.

Even before Chinese public health officials could confirm that the rapidly growing health crisis in the Shanghai area was the likely first sign of an emerging influenza pandemic, cases begin to show up all over the world. Almost all of the early cases had just recently returned from travel to Shanghai and neighboring cities and towns. But that changes quickly when hospitals in other countries receive cases that had never been to China. The WHO, the CDC, and other national health organizations around the world commence their methodical disease detective work. They identify the early cases that presented at each worldwide location and trace their travels back in the weeks before they became ill. Their investigation confirms everyone's worst fear: We are watching the early days of a quickly growing pandemic. No use closing borders; H7N9 has probably taken root in thirty or forty countries by now.

The increasingly nervous experts know that you don't have to touch a sick person to contract seasonal flu as you would have to with Ebola, have sex or exchange bodily fluids as you would have to with AIDS, or get bitten by a mosquito as you would have to with dengue. All you need for transmission is to have someone breathe on you—in a shopping mall, an airplane, a subway, or even a hospital emergency room.

A Middle East terrorist group and a Japanese apocalyptic sect each claim responsibility for the outbreak. The terrorist statement implies that the strain was engineered by former Soviet bioweapon scientists and is a chimera, a combination of the properties of several strains. Both groups promise more engineered outbreaks to come. In response, the director of the CDC and the secretary of Homeland Security say that while investigations are still under way and all threats are being taken seriously, there is no evidence that the H7N9 outbreak is a terrorist action.

By now, the outbreak is universally referred to as the "Shanghai flu," except in China, where it is referred to as the "Western flu." The WHO convenes a group of influenza experts via conference call; this group is known as "the Emergency Committee." After meeting for less than an hour, the committee strongly urges the director-general of the WHO to declare the H7N9 emerging pandemic a Public Health Emergency of International Concern (PHEIC). At a press conference held immediately after the call, she does just that, declaring the situation a global emergency. The press conference turns into a shouting event with reporters demanding to know how the WHO is going to stop the spread of H7N9. There are no satisfactory or comforting answers.

In an impressively short amount of time, working in cooperation with labs in the United States, China, and Britain, the WHO announces that all biological and genetic evidence points to Shanghai as the source of the outbreak, where many millions of chickens are hatched, grown, and consumed each month. Chinese health officials question the findings but say they are cooperating completely with international authorities to curb the spread in China and elsewhere.

Genetic analysis identifies a two-gene reassortment that may be the reason for the sudden human-to-human transmission capability of the virus. The one positive finding is that it is not resistant to current antiviral drugs. The makers of Tamiflu and Relenza go into round-the-clock production but cannot even come close to meeting demand. No vaccine matches this strain, so the US government, working with the WHO, begins developing a vaccine strain of H7N9, to be shared with vaccine manufacturers around the world. The director of the National Institute of Allergy and Infectious Diseases states that he hopes to have an effective vaccine by September or October; that is a long five-plus months away. In less than a week, however, even though the currently available flu vaccine does not protect against H7N9, all available stocks are exhausted.

During an appearance on *Meet the Press,* the CDC director is questioned about H7N9 and asked if it is true that the virus has a 30 percent fatality rate. "While this was true in its limited clusters in China," he replies, "as it disseminates widely, we would expect it to attenuate as it goes through an endless series of human hosts, and the fatality rate should go down considerably."

"Does that mean the deaths we've been seeing from the disease will start to taper off?" the reporter asks.

"I can't say that," the CDC director concedes. "At this point, we still don't know what it is going to do. The best advice I can offer is to try to stay away from those who have influenza-like symptoms. Shelter in place if necessary. And if you have these symptoms yourself, or anyone in your family does, please stay home from work, school, or normal activities where you would interact with other people. Don't travel via public transit either if at all possible; this includes planes, trains, buses, and taxis."

It is now late May, almost six weeks since the newly emerging H7N9 influenza pandemic was recognized in China. At least seventy-two countries are reporting a rapidly increasing number of H7N9 cases and subsequent deaths. The general belief is that more countries have cases but have been reluctant to report them for fear of border closings and trade and travel restrictions. The best data we have on deaths is from the United States, Canada, and the European Union, where case mortality appears to be about 12 percent. So far, at least 12,000 people have died in the United States. Many of the dead are young pregnant women.

Now spot shortages appear in various industries, particularly those impacted by a large disruption in manufacturing in China. It doesn't help that workers at the major seaports and seamen and merchant marines on the 62,000 ocean freighters around the world are reporting an increasing number of ill workers and growing number of deaths. Worldwide, production slows on certain products that have numerous source parts, like computers and automobiles. As news of the epidemic's origin becomes a central part of the international news coverage, consumers are afraid to buy chicken or pork products, regardless of where they came from. Beef prices skyrocket as the supplies tighten.

Doctors' offices and emergency rooms are overrun with the worried well, and the task of physically separating them out from the sick becomes overwhelming. This becomes even more of a challenge as an increasing number of healthcare personnel are too sick to work. Patients demand antibiotic prescriptions even though they are told they are completely useless against viruses. Many who believe they have some knowledge of medicine counter that they want to protect themselves against a secondary bacterial infection. Hospitals are already seeing shortages of critical drugs and supplies. While the US government has a strategic national stockpile for what are called medical countermeasures, or MCM—drugs and supplies needed during a public health emergency—the stockpile is quickly exhausted. Numerous other critical items—enough syringes, needles, antiseptics, diagnostic test kits, and so forth—were never considered for and included in the emergency list.

Some healthcare institutions, like the Mayo Clinic, have planned ahead and at least have a stockpile of Tamiflu that they administer to their physicians and staff, as well as their family members if they develop influenza-like illness. But there is not nearly enough for the patients, including ill healthcare workers, in the developed world countries, and there is virtually none for the rest of the world. Most hospitals are running low or have exhausted their supply of N95 respirators needed to protect healthcare workers. An increasing number of frightened healthcare workers, including both doctors and nurses, are calling in sick. Their illness is fear, not infection.

Virtually every drugstore and pharmacy in the nation has had a run on Tamiflu and Relenza, and there are sporadic reports of break-ins and looting. Most stores have put up signs in their windows declaring that they do not have the drugs. The Internet is flooded with offers for other agents that are effective against H7N9. The commissioner of the Food and Drug Administration warns consumers that there is no evidence that any of these work, and since they are unregulated, they may very well be harmful.

At the direction of the attorney general, the FBI sets up a special task force to investigate allegations of price gouging and black-market sales of antiviral drugs.

On Capitol Hill, chairmen of the relevant oversight committees call on the secretary of HHS and CEOs of the vaccine-manufacturing companies to determine if anything can be done to speed up vaccine production. Other senators and congressmen call for the suspension of flights to and from afflicted countries, only to be countered by experts who say that will no longer make any difference. Some call for trade to be cut off with China, but so many goods and products are already in short supply that this seems to be another useless or counterproductive recommendation.

In Germany, the CEO of one international pharmaceutical corporation is shot outside his home in an apparent assassination attempt, even though his company does not produce vaccines or antivirals. Around the world, other pharmaceutical executives beef up their own security, as fear and frustration turn increasingly to rage and violence.

By early June, the surgeon general has gone on television from the White House to urge anyone who does not need acute care to stay home and not further burden the hospitals. He gives the phone number of a twenty-four-hour hotline where people can consult about their symptoms and see if they need medical or hospital care. Within minutes of the announcement, it is nearly impossible to get through to the hotline. The surgeon general also assures viewers that more Tamiflu and Relenza are in the pipeline, but the public will have to be patient.

Then the president appears, quotes President Franklin Roosevelt, saying, "The only thing we have to fear is fear itself," and decries the recent murders of physicians and pharmacists who were rumored to have supplies of the antiviral drugs.

The lead editorial in the next day's *Wall Street Journal* disagrees with the president, saying, "The only thing we have to fear is a rampant and deadly influenza epidemic for which this country was totally unprepared and which this administration has been far too slow to respond to." The editorial traces the 50 percent decline in American stocks since the beginning of the pandemic, with commensurate drops around the world, and the near collapse of the Chinese exchanges.

Attendance plummets at sporting events, theme parks, and shopping malls. Most public events are now canceled. Major League Baseball is considering temporarily suspending its season. Retailers and park operators have to lay off large percentages of their already diminished workforce. National unemployment soars above 25 percent, while certain industries can't find enough qualified workers. Many automobile dealers are now open only on weekends for new car sales, and their service bays are nearly empty. The Federal Reserve lowers the federal funds rate to zero.

Huge poultry farms are culled in Shanghai and Hong Kong, and producers worldwide say there is no reason to build up their stocks again until the pandemic is over, since consumption has tanked. Food supplies are getting tighter and tighter worldwide, even on the grocery store shelves of America.

Although some small towns and rural areas have been largely spared from the infectious scourge, by June a national survey shows that most people say they know someone who has died from the Shanghai flu. Several newspapers have taken to running a photo spread each week of local residents who have perished.

The president appoints a Shanghai flu czar to head a task force made up of the heads of virtually every possible federal government agency with an interest in vaccines, public health, and emergency preparedness. The American manufacturers predict they will be able to produce a steady supply of vaccine beginning in late September, but altogether this will cover no more than 40 percent of the population for the following five months. No other nation will commit to sending any of their supplies to the United States, given that they are in the same position. The two countries with large production capacities—India and China—say that they can cover no more than 10 to 15 percent of their own populations. Early batches of vaccine from one Indian manufacturer turn out to be contaminated with a bacterium and must be thrown out. Everyone begins to realize that most of the world's population will never have an opportunity to get vaccinated for H7N9. And the question about how well the vaccine works in protecting people from H7N9 infection has not been answered, but it is the only vaccine available.

By the first week of July the casualty rate has started to decline. Within weeks, hospitals are recording only a few new cases. The CDC reports that though there are sporadic hot spots around the world, the flu appears to be abating. The stock market starts to climb, while analysts warn that this may only last until earnings season, when we will see how much damage the pandemic has done. The loss to worldwide gross national product is difficult to measure, but it is certainly in the many trillions of dollars. Everyone says it will take years to recover.

The CDC estimates the total number of cases in the United States at 31 million, or approximately 9 percent of the population. Of those, deaths totaled approximately 1,932,000, for a fatality rate of around 6 percent. Global statistics are not yet available but are thought to be at least as severe.

The president proposes August 1 as a day of public reflection and personal commitment, as well as a celebration of the fact that the nation and most of the world has survived its greatest challenge since World War II. This ordeal has been a message that we all have to pledge ourselves to the common good. We should use both the many examples of great heroism and personal sacrifice and the instances of greed and incredible selfishness during the crisis as a moral compass going forward.

Public health leaders urge the president to postpone such a celebration. They warn that based on the history of previous pandemics, a likely second wave of illness could start in the early fall and actually exceed the number of cases and deaths that occurred in the first. Like the first wave, a second wave could last ten to twelve weeks in the United States, or even longer. They say it was unfortunate that the world needed so deadly a wake-up call to take seriously the impact of the influenza pandemic they had been predicting for so long.

Influenza news slowly disappears from television and is relegated to the back pages of newspapers. When the epidemic is mentioned, it is usually in terms of "economic recovery from the Shanghai flu pandemic."

It is late September when new cases start showing up at physicians' offices and hospital emergency rooms. The antigen tests quickly confirm H7N9 influenza virus, meaning that the outbreaks earlier that month in Cairo, Egypt, and Lahore, Pakistan, were not flukes.

A series of conference calls is launched by the White House, including the "kitchen sink" of federal, state, and local agencies such as HHS, the CDC, the NIH, the Public Health Service, the FDA, the Department of Defense, the Department of Homeland Security (including the Federal Emergency Management Agency), and state health and emergency preparedness agencies, to organize and coordinate plans to get the new Shanghai flu vaccine distributed throughout the country. It is anticipated that the first vaccine will become available the last week of September in the United States and Canada, and the following week in Great Britain and parts of the EU. The first vaccine will go to health-care workers, first responders, and critical government employees such as firefighters and police. There is a huge outcry from the public that doctors and nurses and the government are just taking care of their own. The argument is made by federal health officials that if these individuals are not protected, more people will die due to lack of healthcare workers and emergency response. When the first vaccine does arrive in each state, clinics are set up in hospitals for healthcare workers and others in the critical vaccination group, together totaling more than 25 million people. But word leaks out about when and where these vaccine clinics are being held, and they are overrun by masses of people seeking vaccination. Chaos prevails. Police, who are already short staffed because of cases in their own ranks, try to protect the vaccinators and the vaccines. Outbreaks of violence at these clinics are reported throughout the United States.

The US vaccine supply will continue to increase by late October, but it is unclear how much will be available, and it will be far less than needed. Anticipating the new stocks, government officials decide that large parking facilities, shopping centers, and stadiums will provide the best venues for vaccination. All sites will be supported by state and local police units.

Despite these precautions, when vaccine does arrive, many locations are overrun by masses of people, and when the supplies are quickly exhausted, the crowds turn violent. Though no one is killed, there are numerous injuries.

The WHO's director general, who five months before had declared a Public Health Emergency of International Concern, has no advice to offer other than to try to stay away from infected individuals. Surveillance suggests a mortality rate between 4 and 6 percent in those who contract the Shanghai flu in Western nations, but it is considerably higher in developing nations, where the healthcare systems have completely broken down. In addition to the influenza deaths, mortality from all other sources has doubled. In Central Africa, vaccine-preventable childhood diseases and TB are said to be out of control because of a lack of basic medical care and public health services.

Hospitals in the United States suffer another round of severe product shortages. They first experience a shortage of saline bags and disposable syringes, but soon supplies of basic lifesaving drugs dwindle. The American Diabetes Association warns for a second time in four months that unless insulin stocks are resupplied soon, people will die. Most hospitals curtail all elective surgeries until further notice. All mechanical ventilators in the United States are in use, but they can treat only a small minority of those who need them. Many others die, particularly the elderly. Again, healthy men and women in the prime of life suffer exaggerated immune system reactions. Pregnant women are especially vulnerable. As with the Zika virus outbreak, health authorities around the world recommend that women of childbearing age postpone pregnancy.

Food shortages happen even faster this time. Because of the run on food stores when the second wave was announced, shelves are largely bare, particularly of meats, dairy products, produce, and other perishables. Many stores close rather than risk looting or vandalism. There is little violence against drugstores this time, though, because it is common knowledge that they have no vaccine or critical pharmaceuticals.

However, virtually all governors have called out the National Guard to quell the riots and large demonstrations protesting the lack of vaccine, antivirals, and other medical support. This time, a special federal court is established to deal with accusations of profiteering, black marketing, and phony drugs and medical supplies. In China and several African and Middle Eastern countries, offenders are publicly executed.

When it is announced that the absentee rate due to the influenza is approaching 30 percent, there is fierce debate in Congress and in the media on whether to allow Mexican seasonal workers into the country to harvest crops. Conservative lawmakers worry that they will bring even more disease with them. The NIH director is called before the US Senate Committee on Health, Education, Labor and Pensions. The committee chairman reads statements in which the director has repeatedly predicted over the past five years that a universal influenza vaccine will be forthcoming, yet none exists. The director mumbles something about funding and commitment but has no real response.

In New York, the subway system has virtually shut down as commuters realize they cannot avoid being breathed on. The streets are essentially gridlocked with private cars. The director of the Environmental Protection Agency warns of dangerous levels of air pollution. It is difficult to estimate the daily loss of productivity, but it is clearly in the tens of millions of dollars.

The world's stock exchanges, which had been gradually creeping up since July, plunge again, giving up another large percentage of their already anemic value. The gross national products of all developed nations have decreased by nearly half and the world is officially in economic depression. The American unemployment rate reaches 22 percent—less than three points below that of 1933: the worst year of the Great Depression.

By now, almost every major city around the world is witnessing people dying in offices, in public buildings, and right on the streets. Morgues are overflowing with bodies and there is a worldwide shortage of coffins. Developing countries begin cremating corpses in large ditches that are then immediately covered over by bulldozers. In the United States and other First World nations, morgues are forced to supplement with freezer trucks, but the spot shortages of electricity and fuel are forcing some difficult decisions on disposal.

Certain right-wing televangelists state that Shanghai flu is God's punishment for straying from his ways. Public health leaders condemn this "dangerous and irresponsible fearmongering that can only distract us from our real challenges." They emphasize that "no one is responsible for becoming ill, but all should take whatever precautions they can."

The American president and the other leaders of the G7 nations meet via secure video link because of the concern about travel. They release a statement that the H7N9 pandemic "is the moral equivalent of war," with all the world's people engaged together in a mortal battle with a common enemy deadlier than any human adversary.

In most places, panic and civil strife have now given way to an overriding sense of resignation. Streets of major cities are close to empty. Stores, restaurants, and entertainment venues are closed. Researchers are more certain how the H7N9 changed into the pandemic strain, but to most of the public, the question seems largely academic. Vaccine stocks continue to trickle in and are quickly used, but so many people have suffered or died from the illness that demand is actually beginning to drop.

By the following June, when the pandemic has finally run its primary course, the worldwide death toll from the two disease waves is approximately 360 million, out of nearly 2.22 billion total cases. The average age of those who died is thirty-seven. While the percentage of those who have died around the world does not come close to that of those who died in the Black Death, which wiped out nearly a third of the population of Europe and the Mediterranean region in the fourteenth century, in terms of raw morbidity and mortality statistics, the Shanghai influenza pandemic is by far the largest catastrophe in world history.

The preceding scenario is fictional but far from fanciful.

On May 10, 2016, the National Health and Family Planning Commission of China notified the WHO of eleven new cases of laboratory-confirmed human infection with H7N9 influenza. Four of the patients had died and two more were in critical condition at the time of the report. The two in critical condition—a twenty-three-year-old male and a forty-three-year-old female—had exposure to each other. Therefore, the WHO noted, "Human to human transmission between the two patients cannot be ruled out."

According to the WHO's risk assessment statement, "Since the virus continues to be detected in animals and environments, further human cases can be expected." Then, a few sentences later: "Human infections with the A(H7N9) virus are unusual and need to be monitored closely in order to identify changes in the virus and/or its transmission to humans as it may have a serious public health impact."

There is no way to know how many warnings we will get before the events we have portrayed here become all too possible. They may not be far off.

Few people see this more clearly than Ron Klain, who oversaw our international response to the Ebola outbreak in West Africa:

> If my experience coordinating our Ebola response did not make me an infectious disease expert, it did give me a battlefield expertise in what works—and what does not—in our global policy and governmental frameworks in responding to an infectious disease outbreak and epidemic. And it left me with the perspective that while we did make some progress in preparedness—as a country and as a global community—during the Ebola epidemic, I'm sad to say that as we stand here today, the world still has gaping holes and glaring inadequacies in its preparedness for a ghastly eventuality that is certain to come. Those gaps are not only in poorer countries with weaker medical systems, as one might expect, but even here in the United States, with our envy of the world institutions and resources.
>
> Why is this so worrisome? Because it seems likely that the world is living on borrowed time before one of these new infectious disease threats becomes the kind of global pandemic we have all been warned to expect. It is not hard to

imagine that some time during the next president's term, his or her national security team may be summoned to the Oval Office to discuss a catastrophic pandemic of historic proportions: more than one million deaths in just a few weeks in a far corner of the world, sparking the fall of several governments, giving rise to a violent regional conflict over scarce resources, and unleashing a refugee crisis as fleeing victims encounter panic and closed borders at every turn. Worse still, the president will be told, there is an increasing risk that such death and disruption may soon arrive in the United States.

CHAPTER 20

Taking Influenza off the Table

A pessimist sees the difficulty in every opportunity; an optimist sees the opportunity in every difficulty.

— Sir Winston Churchill

Our current influenza vaccine is unique, and not in a good way.

As we have noted, influenza is the only disease for which a vaccine has to be administered each year. This is because the HA and NA antigens drift so rapidly that the antibodies developed by our immune systems from a previous exposure to either vaccine or the actual virus can't recognize new influenza viruses. This new yearly vaccine is based on worldwide, less-than-foolproof surveillance, leading to a collective guess on which strains will be dominant the following fall, winter, and spring; and the vaccine is developed and manufactured largely with technology that is now more than sixty years old. Even when we get the virus match right, protection may be limited for reasons we don't fully understand.

It was in 1933—more than twelve years after the end of the 1918 pandemic—that Dr. Richard E. Shope of the Rockefeller Institute laboratory in Princeton, New Jersey, identified influenza as a virus by transmitting it in a fluid among pigs and passing that fluid through filters too small to pass either bacteria or fungi. Since then, the race has been on to come up with an effective vaccine.

Think of the HA antigen like a stalk of broccoli, where the head sticks out on the surface of the virus and is frequently changing its structure. Meanwhile, the HA stem is buried in the virus and rarely changes. This is an important observation, as we have increasing evidence that making an immune response to the HA stem may be broadly protective against multiple strains of influenza viruses.

Most influenza vaccine, even with improvements in manufacturing techniques, takes six to eight months to produce and is grown in pathogen-free embryonated chicken eggs (meaning they have embryos). Few people know that we maintain a strategic stock of chickens for this purpose, since you need a whole lot of eggs to produce sufficient stocks of vaccine. Some influenza vaccine is now grown in cell culture, but it, too, can take months to produce.

The most significant drawback of the cell culture method is that it still doesn't produce vaccines that are more effective than those grown in chicken eggs. In fact, flu vaccine is one of the poorest-performing vaccines in our medical armamentarium. Is it better than nothing? Generally so, but in some years by no more than 10 to 40 percent.

In October 2011 our group at CIDRAP and colleagues at the Marshfield Clinic and the Johns Hopkins Bloomberg School of Public Health published a paper in the medical journal *Lancet Infectious Diseases.* We showed that since the mid-1940s, when influenza vaccination became widely available, most studies of its efficacy have relied on suboptimal methodology, and the actual protection offered has been significantly lower than the medical community and the public believed. This has been particularly true for individuals over sixty-five years of age—the cohort of the population most vulnerable to seasonal influenza.

We have too few good studies to determine the effectiveness in older people, but we found, on average, the vaccine works about 59 percent of the time in protecting younger adults. In some years it's much less effective than that. For example, for the H3N2 strain, the 2014–15 vaccine actually provided 0 percent protection.

When we published this paper we were taking on one of the sacred cows of public health: the long-held belief that seasonal influenza vaccine protects 70 to 90 percent of those vaccinated. These were the numbers the CDC and other public health and medical organizations had been actively promoting for years. I received some pretty unpleasant e-mails and phone calls from public health and medical colleagues after our article was published. Some even compared me to Andrew Wakefield, the British physician who put forth false data to show that measles vaccine caused autism—though it does not. It was not a pleasant time for our group, but we knew we were right. In fact, it has been this sloppy science and subsequent promotion of our current influenza vaccines that has held us back for many years from fully realizing why we must have significantly better vaccines.

Tony Fauci is adamant about what we have to do in this regard. "We need to realize right now that we do not have an adequate vaccine for influenza," he told us. "And we need to figure it out in the same way that we're putting an incredible amount of money into trying to figure out if we can get a vaccine against HIV. I think we were lulled into some kind of complacency, because we had an influenza vaccine that we essentially used every year, that we modified a little bit to account for drift and shift. And we never said, 'Wait a minute; we've got to do better than this!'"

Influenza vaccine policy for the last fifteen years or so, both in the United States and internationally, has been focused on ensuring that capacity exists to produce enough seasonal vaccine so that larger and larger segments of the population can be vaccinated, particularly in developing nations. This approach was supported by both government public health agencies and a vaccine industry that counts on having a stable market in which to sell vaccine and thus realize a steady annual profit. While these goals are important interim measures given the present landscape of influenza vaccine science, they are not sufficient to address the big-picture challenge. That is: Public health policy experts and the vaccine industry have not focused on the limitations of the current vaccines that target antigens in the changeable HA head.

For example, when the federal government did an exhaustive review of the 2009 H1N1 pandemic vaccine response, it never asked how well the vaccine actually protected, simply whether it was available in time for the second wave, which it largely was not. In fact, in a well-done study by the CDC, the overall protection of the vaccine was shown to be only 56 percent. How that fact could be omitted from the US government's review is beyond me. The current general policy approach to improving the vaccine is to make incremental changes to existing HA-head vaccines. These efforts might lead to some improvements, but the overall impact will be small.

Since our 2011 *Lancet Infectious Diseases* paper, a series of annual-vaccine effectiveness studies has been conducted in the United States, Canada, Europe, and Australia. Most of these studies have been supported by the CDC and use methods that avoid the problems of the previous studies. Their results fully support our conclusion about the variable vaccine protection each year and the far-less-than-optimum effectiveness in most years. There are also several new studies that suggest it is actually better *not to have* the vaccination each year, that such a practice may actually cut down on antibody response. More investigation is needed to show whether, in fact, this holds across the age and health spectrum and, if so, what is the most effective interval between seasonal flu shots or mists. At this point we should be honest enough to admit that we just don't know.

In October 2012, CIDRAP published the detailed report referenced in chapter 10 on vaccines: *The Compelling Need for Game-Changing Influenza Vaccines: An Analysis of the Influenza Vaccine Enterprise and Recommendations for the Future.* We refer to the report as CCIVI, for CIDRAP Comprehensive Influenza Vaccine Initiative, and I believe this work remains the most comprehensive cradle-to-grave analysis ever undertaken on any vaccine.

In the CCIVI report, we covered everything from an overview of influenza infection to current licensed vaccines, safety, public acceptance, vaccine availability, influenza immunology, potential game-changing vaccines in the research pipeline, regulation, financial and market considerations, and public health policy, organization, and leadership barriers.

We identified four reasons for our collective failure to secure twenty-first-century influenza vaccines. First, for several decades, public health was our own worst enemy in making the case for the urgent need for new influenza vaccines. Because we incorrectly told the world that this vaccine was 70 to 90 percent effective, policy makers, vaccine manufacturers, and investors have had little interest in finding new and improved vaccines. Second, because there has been only limited public investment in researching new influenza vaccines, we still lack a level of research and development necessary to bring new vaccines forward through the investigation and licensing process. Third, a sound business pathway must be identified that will overcome the financial disincentives of the current influenza vaccine manufacturers to end their annual vaccine sales market and adopt a market where a vaccine may be administered only once every decade. If the industry is not on board, no one will make these future vaccines. Finally, nobody is in charge of making these new influenza vaccines a reality; not governments, industry, academia, or organizations like the WHO. When I attend meetings with leaders from these groups, we all agree that there is an urgent need for these new influenza vaccines, yet everyone points the finger at someone else as needing to be in charge of making it happen. Government agencies point to the vaccine industry as the needed leaders, and in turn, the industry states the government should take the lead. I even found the same problem among the participants in the Coalition for Epidemic Preparedness Innovations regarding influenza vaccines. The conclusion of this group was that we should not take on the task of supporting new influenza vaccines because the industry is already doing that—but they are not in any meaningful way. Until these issues are addressed and answered, new influenza vaccines are going nowhere.

The preceding chapter, we think, makes the case for what would happen if we simply sat on our hands on this issue and did not come up with significant improvement in our current influenza defenses. But let's hear from someone on the inside.

Stewart Simonson is someone on the inside who served Governor Tommy Thompson as chief legal counsel and followed him to Amtrak and then the Department of Health and Human Services. Simonson joined HHS a month before the 9/11 attacks, and from then on coordinated the department's efforts on biodefense and public health preparedness. In 2004 he became the first assistant secretary for public health emergency preparedness and continued in that position under Thompson's successor, Mike Leavitt. In that position, he impressed me deeply with his dedication, comprehension of the subject, and creative imagination in how to make the government effective in emergency preparedness.

When we asked him about an influenza pandemic sometime in the indeterminate future, and how well prepared we are, he replied, "We know that influenza can cause a catastrophe. We know it because it has happened, and will happen again: that which is not prohibited is compulsory." This is a play off a famous quote from T. H. White's *The Once and Future King*, and to me it means that if something is possible, in our kind of planning, it is inevitable.

"It's not a low probability," Simonson continued.

> It is a high-probability, low-frequency threat. So it will happen; that is a given. The variables are when and how severe; and, of course, how prepared mankind will be to respond. As you know, Mother Nature is the greatest bioterrorist of them all, with no financial limitations or ethical compunctions—at least that we understand—and no limit on the level of effort expended. Our most dangerous adversary will not originate in the tribal areas of Afghanistan or some other remote place. It is everywhere man and animal live in close proximity. Just ask the chickens. As we used to say at HHS: If you're a chicken, it's already a pandemic.
>
> And for something like this, you can't turn on a dime. You have to have a ten-year runway. The problem is, with any of these threats, Congress gets worried, they appropriate a lot of money. What isn't obligated gets taken and put into the next threat, and the next.

There can be no greater bang for the buck than investing in what I call a game-changing influenza vaccine. In any given year, or even in any given decade, the probability of a major influenza pandemic is low. As a possibility for some unknown point in the future, it is virtually a dead certainty.

What do we mean by "game changing"? Many in the public health community talk about a "universal" influenza vaccine that theoretically could, as we explained in chapter 8, target those elements of the virus that remain the same in all strains. I believe this is an unrealistic goal, both scientifically and economically. But we can get close enough.

Remember from chapter 19 that influenza A can have one of eighteen different HAs and eleven different NAs. Human disease is caused primarily by HA 1, 2, 3, 5, 7, and 9 and by NA 1, 2, and 9. If we can develop vaccines that protect against just the six HA and three NA types that currently infect humans, allowing for the possibility of new HA and NA strains emerging, even if antigenic drift and shift occur in the virus, we will have a vaccine that could essentially take pandemic influenza off the table. And that would certainly "change the game."

"Once you do that," says Tony Fauci, "you take a different kind of an approach. Likely what will happen if we do it right is that we'll get something close to what we're hypothesizing now. There is no reason at all not to have long-term [antigenic] memory for influenza if you get the right induction of the right immunogens. So I think we need to re-look at the entire influenza theme."

We also want a vaccine to protect us for a number of years after receiving a single dose, rather than having to get shots each year. I believe such a vaccine is within our reach. Remember, I'm the guy who in 1984 said I didn't think I'd see an effective HIV vaccine in my professional lifetime, so you can't say I'm an irrational optimist.

We would want this game-changing vaccine to be produced with manufacturing techniques that could easily be scaled up and used as part of an ongoing global campaign against seasonal influenza, to make the possibility of a global pandemic much more remote.

In the CCIVI report we detail other attributes of a game-changing vaccine that would be helpful. It must be cost-effective enough to be distributed worldwide, as are childhood immunizations; the manufacturing techniques to make the vaccine should be readily transferrable to nations of the developing world; it should be heat stable so a "cold chain" is not necessary to transport it from the factory to the field destination; and, if possible, it would not require an injection at all, but could be administered through some more efficient and less invasive means.

Is this realistic, or wishful science fiction?

"We need to really probe the science," says Tony Fauci. "It isn't an engineering problem; it's a science problem. So we just need to crack that. It's going to take a major effort, the same as we're doing with HIV."

Though in science, proof of concept doesn't always translate into proof of effectiveness, there are several promising technologies currently at the experimental stage. None is dependent on the creaky, decades-old process involving chicken eggs.

Immune response results in these initial game-changing influenza vaccine studies have been mixed, and there are many hurdles still to be surmounted. From 2007 to 2014, I directed the Minnesota Center of Excellence for Influenza Research and Surveillance, one of five major NIH centers doing influenza research. I'm still a researcher in this effort, and some of the best minds in the influenza immunology business are coinvestigators in this network. They don't minimize the challenges of finding a game-changing influenza vaccine, but they do believe it's doable. The biggest holdup we have in moving forward is a lack of coordinated leadership and sustained adequate funding.

The pathway to licensure for these vaccines will be complicated. Large randomized, controlled efficacy trials will be required. Because these new vaccines will not be based on generating antibodies to the HA head, as previous vaccines have been, new immunological metrics will have to be developed and assessed.

As of today, there are nineteen potential game-changing

influenza vaccines in either Phase I or Phase II trial status with the FDA. I realize that some of these candidates may be seen as too risky to invest a billion dollars in a Phase III trial, but the only way we're going to achieve a game-changing vaccine is to get something workable past the valley of death.

In a way, this is like saying we have developed a prototype for a new, highly efficient supersonic airliner. The only problem is we can't get it off the ground to test it because no one has built the runways to allow it to take off.

As we suggested with the development of new antibiotics and other antimicrobial agents, if we are to come up with a game-changing vaccine that will essentially take influenza off the table as a global concern, we can't expect private industry to carry the burden alone.

In addition to all of the developmental and clinical costs, a game-changing flu vaccine would change the current business model that relies on selling new vaccine doses each year. With the new game-changing vaccine, we will, hopefully, have to vaccinate people only once every decade. In a typical seasonal flu year, the global vaccine market is close to $3 billion. This figure would be several times larger during a pandemic, even a relatively mild one. But with a game-changing vaccine, once the manufacturer gets past the initial surge of sales in countries like the United States, Canada, and those in Europe, there are 6 billion–plus more people in the rest of the world, and the more of those we can vaccinate, the lower the risk of another pandemic.

If the vaccine industry doesn't see the possibility of a global market for a game-changing vaccine, it is highly unlikely this vaccine will ever see the light of day, unless there are major government- or foundation-provided incentives. While we have seen many policy documents that recognize the need to develop game-changing influenza vaccines using new approaches and technologies, the political will to provide the resources and strategies necessary to make any of them a reality has been almost completely lacking.

What we propose, therefore, is to implement a Manhattan Project type of effort, after first launching education and outreach similar to that which preceded the NASA space program to make the public aware of what a tremendous benefit this would be to all of humankind. If we could get across the idea that a game-changing influenza vaccine could be as impactful as the smallpox vaccine, we believe the cost and value of the program would be an easy sell.

The Manhattan Project, as most people know, was the American government's urgent secret program to research, develop, and test an atomic weapon. Only our program, to create a game-changing influenza vaccine, would not have to be secret. The term "Manhattan Project" has become synonymous with an endeavor of great effort, expertise, and resources brought together to achieve a specific objective, and the project has been widely recognized as one of the most successful project management efforts of modern times. At its height in 1944, it employed 129,000 workers of all sorts, involved major construction at ten different sites in three countries, and cost more than $2 billion—close to $30 billion in today's dollars.

After studying the many scientific, logistic, legal, procurement, public and private partnership relations, resource priorities, and management requirements involved in a universal influenza vaccine development enterprise, we believe that the Manhattan Project serves as a relevant and useful model. First, the project was determined to be mission critical by the highest levels of the US government. Second, it was resourced accordingly. Third, the best principles of project management were employed to complete the mission in a secure and timely manner.

One could even consider a model like the International AIDS Vaccine Initiative, known as IAVI. It is a global not-for-profit public-private partnership working to accelerate the development of vaccines to prevent HIV infection and AIDS. IAVI, with an annual budget of more than $1 billion, researches and develops vaccine candidates, conducts policy analyses, serves as an advocate for the HIV prevention field, and engages communities in the trial process and AIDS vaccine education. IAVI's scientific team comes from private industry and more than fifty academic, biotechnology, pharmaceutical, and governmental institutions. The major donors to IAVI include twelve government or multinational organizations, thirteen foundations, and twelve companies.

Our best estimate today is that only $35 to $40 million of public and industry support globally is being spent on researching game-changing influenza vaccines. This investment pales in comparison to the $1 billion spent annually on HIV vaccine. Imagine what we could do if research on a game-changing influenza vaccine was funded at a similar level to HIV and done in a coordinated and collaborative manner.

We recognize the current environment of fiscal austerity. However, as we have shown, the social, economic, and political consequences of a severe influenza pandemic on the entire world in the absence of a readily available and effective vaccine cannot be overstated. Our ultimate goal should be to have a dose of game-changing influenza vaccine for every human being on earth.

The London-based worldwide professional services company Willis Towers Watson polls 3,000 insurance industry executives each year for what they consider the greatest risks to their industry, in other words, what would cost them the most. We looked at the *Extreme Risks* survey for 2013. Number three on the fifty-seven-place ranking is "Food/water/energy crisis: A major shortfall in the supply of, or access to, food/water/energy, causing severe societal shortages." Number two is "Natural catastrophe: A confluence of major earthquakes, tsunamis, hurricanes, flooding and/or volcanic eruptions with major global effects."

At the top of the list is "Pandemic: A new, highly infectious and fatal disease spreads through human, animal or plant populations worldwide."

That pandemic is most likely to come in the form of a deadly influenza strain.

CHAPTER 21

Battle Plan for Survival

"Before I draw nearer to that stone to which you point," said Scrooge, "answer me one question. Are these the shadows of the things that Will be, or are they shadows of things that May be, only?"

Still the Ghost pointed downward to the grave by which it stood.

"Men's courses will foreshadow certain ends, to which, if persevered in, they must lead," said Scrooge. "But if the courses be departed from, the ends will change. Say it is thus with what you show me!"

— Charles Dickens, *A Christmas Carol*

We have no illusions about what is likely to be accomplished on our Crisis Agenda in a world divided on so many levels. But we also have no illusions about what *must* be done if we are to make our world a safer and healthier place for our children and grandchildren, where pandemics do not threaten our way of life on every level imaginable, where infections caused by drug-resistant microbes do not kill for lack of an effective treatment, where drinking water does not become a vehicle of death, and where the emergence of new infectious diseases does not become a public health crisis because we are not prepared to rapidly stop them. If we do not do what we collectively need to do, the shadows of things that *may* be are almost certain to become the harsh reality of what *will* be.

With this book, we have intended to present the face of infectious disease in the modern world. We have tried to connect as many dots as possible, especially from science to policy. Moving toward our conclusion, we've surveyed the ideas and observations of some of the best minds in public health and public policy. I have used all the lessons that I have learned from my forty-plus years of fighting to prevent and control infectious diseases. This final chapter lays out, in order of priority, what we must do to alter the otherwise catastrophic potential of infectious diseases on humans and animals.

To review, our greatest threats are:

1. Pathogens of pandemic potential, which essentially means influenza and the downstream effects of antimicrobial resistance.
2. Pathogens of critical regional importance, which include Ebola, coronaviruses like SARS and MERS, other viruses such as Lassa and Nipah, and *Aedes*-transmitted diseases such as dengue, yellow fever, and Zika.
3. Bioterrorism and dual-use research of concern (DURC), and gain-of-function research of concern (GOFRC).
4. Endemic diseases that continue to have a major impact on the world's health, particularly among emerging nations, including malaria, tuberculosis, AIDS, viral hepatitis, childhood diarrheal diseases, and bacterial pneumonia.

We must consider these threats within the context of certain factors. The most critical of these are climate change, availability of water for drinking and irrigation, global governance and fragile state status, economic disparity, and the ongoing struggle to empower women.

We address these four threats with a nine-point Crisis Agenda. We provide specific program recommendations that largely have not been addressed by the federal government, public health organizations, or even recent formal reviews of the global public health response to the West African Ebola epidemic.

These priorities are listed in order of importance, that is, their potential impact on overall global public health and early, avoidable death.

The Crisis Agenda

Priority 1: Create a Manhattan Project–like program to secure a game-changing influenza vaccine and vaccinate the world.

The single most consequential action that we can take to limit, and possibly even prevent, a catastrophic global influenza pandemic is to develop a game-changing influenza vaccine and vaccinate the world's population. This is scientifically attainable, though the CCIVI report concluded that only the US government has the necessary infrastructure and resources. We need only the creative imagination of our best scientists, the visionary support of our policy leaders, the technological and financial commitment, and the necessary project-management structure. We would hope other national governments, philanthropic organizations, vaccine manufacturers, and the WHO would readily join the effort. Our best guess is that we would need to invest $1 billion per year for seven to ten years to make this happen. This is about what we currently invest each year in HIV vaccine research, and I believe we'd have a greater chance of the influenza vaccine working. Vaccinating most of the world before another catastrophic pandemic has a chance to begin could save more lives in just a few months than all the emergency rooms in the United States have done in the last fifty years.

Priority 2: Establish an international organization to urgently address all aspects of antimicrobial resistance.

The Intergovernmental Panel on Climate Change (IPCC) was created in 1988 by the World Meteorological Organization and the United Nations Environment Program "to prepare, based on available scientific information, assessments on all aspects of climate change and its impacts, with a view of formulating realistic response strategies." Since then, the IPCC has served as the scientific authority and moral conscience of all aspects of climate change capably. We must have a similar model for antimicrobial resistance. Like climate change, it is a global crisis in that no one country or region of the world can solve it. And like the greenhouse gases that settle in the atmosphere around the entire planet no matter where they originate, antimicrobial-resistant viruses, bacteria, and parasites will spread around the world no matter where they evolve. The establishment of a panel like the IPCC, under UN authority, will require support and resources from the developed countries to effectively counter the issue of antimicrobial resistance.

Priority 3: Support and substantially expand the mission and scope of the Coalition for Epidemic Preparedness and Innovations (CEPI) to fast-track comprehensive public-private vaccine research, development, manufacturing, and distribution for diseases of current or potential critical regional importance.

The urgent need for vaccines to protect against pathogens of critical regional importance should be obvious. What hasn't been obvious to those outside a small group of public health professionals and vaccine industry experts is that the international system for researching, developing, and distributing these vaccines is broken and desperately near collapse. We should be far beyond the debate about why governments and philanthropic organizations must provide substantial support to private pharmaceutical companies to have these vaccines when and where we need them.

CEPI represents the first real advance in securing such vaccines. It is a novel partnership of the governments of the United States, the European Union, India, the Gates Foundation, the Wellcome Trust, Gavi: The Vaccine Alliance, the World Economic Forum, and leading vaccine manufacturers. Aside from its EU connection, Norway has its own separate partnership with CEPI.

My biggest concern is that CEPI is not thinking big enough. Annual funding under consideration for the first several years is in the range of $200 million. When I look at the portfolio of critically needed vaccines and the resources that will be required to bring them to licensure, purchase, and distribution, I believe an annual $1 billion infusion of support will provide a huge return on investment in terms of both lives saved and direct and indirect economic costs. All the parties are at the table to make this happen. It will be up to them to embrace and support this more aggressive approach. Once we have these vaccines, we need to use them in advance of potentially devastating epidemics. This is where Gavi and the WHO need to step forward to extend the CEPI mission. Imagine if we could launch a massive Ebola vaccine campaign today targeting all those in Africa at potential risk, including healthcare workers, ambulance drivers, public safety workers, and burial team members. Or how about vaccinating healthcare workers and camel herders on the Arabian Peninsula against MERS? In both examples, we may be able to stop emerging, large outbreaks from ever occurring.

While we are addressing the lack of critical vaccines, we also need to take on the lack of critical diagnostic tests, particularly for those infectious diseases that can cause sudden, regional epidemics. Diagnostic tests, especially those that can be done quickly and reliably at the patient bedside, are necessary for recognizing and controlling outbreaks of infectious disease. For example, our inability to reliably and quickly diagnose patients with Ebola infection in West Africa was a contributing factor in the rapid spread of the virus. Unless there is a near-term financial incentive for diagnostic test research and development companies to create and market tests for Ebola, Zika, or other possible agents that might emerge one day, such tests won't be available for the next crisis. We need a comprehensive international CEPI-like initiative to address this major shortcoming if we are to improve our public health and medical care aspects of emerging infections.

Priority 4: Launch the Global Alliance for Control of Aedes-Transmitted Diseases (GAAD) and coordinate with the Bill & Melinda Gates Foundation's malaria strategy, "Accelerate to Zero."

There is urgent need to bring mosquito-control science and practice into the twenty-first century. The past forty years have seen the dramatic emergence of epidemic arboviral diseases transmitted by *Ae. aegypti*. During that same time, the prior high level of investment in, and commitment to, *Aedes*-related control research and professional training has virtually disappeared. There is an immediate need for experts in mosquito-control science and policy to develop an effective overall strategy for *Aedes*-control tools and begin to research new ones such as pesticides. To provide this leadership, world experts in *Aedes* biology and control have proposed the creation of a global alliance of international institutions with a vested interest in preventing *Aedes*-transmitted diseases, to be known as the Global Alliance for Control of *Aedes*-Transmitted Diseases (GAAD). Members would include national governments, nongovernmental organizations, international funding agencies, and foundations. The alliance would be established under a charter with a committee consisting of representatives from each member organization.

A coordinated source of funding would be needed to develop, manage, and implement the program. We believe an initial investment of $100 million annually would be effective. The US government should lead the way with this support, with other countries in the "*Aedes* belt" also making sizable investments. GAAD would need to coordinate its activities closely with the WHO; however, as noted previously, the WHO has no major vector-borne-disease resources or expertise.

The Gates Foundation has already launched a major initiative called "Accelerate to Zero" against malaria, a disease transmitted by the *Anopheles* mosquito. To date, its results have been impressive. While the biology of *Aedes* and *Anopheles* mosquitoes, and thus subsequent control measures, is quite different, coordination of the GAAD and Gates Foundation activities would capitalize on shared research activities such as the development of new, effective, and safe pesticides.

Priority 5: Fully implement the recommendations of the bipartisan report of the Blue Ribbon Study Panel on Biodefense.

The October 2015 report is a landmark document that provides the road map for what we must do to maximize our preparedness for a bioterrorist attack, in the United States or elsewhere in the world. It concludes, "The United States is underprepared for biological threats. Nation states and unaffiliated terrorists (via biological terrorism) and nature itself (via emerging and reemerging infectious diseases) threaten us. While biological events may be inevitable, their level of impact on our country is not."

Today, I'm afraid, the report is accumulating dust on the shelves of the Washington bureaucracy. The next administration and Congress should rank the implementation of the report's thirty-three recommendations of the highest priority.

As former secretary of the navy Richard Danzig told the panel, "We don't really get to choose what we have to prepare for."

Priority 6: Establish an international organization similar to the National Scientific Advisory Board for Biosecurity (NSABB) to minimize the use of DURC and GOFRC to transmit pathogens of pandemic potential.

While we have been critical of the accomplishments of the NSABB, it is nonetheless leading the world in addressing the current and future challenges of dual-use research of concern and gain-of-function research of concern. It is my hope that the NSABB can take the next step and follow through on the recommendations made in chapter 10 regarding additional issues they must address. Meanwhile DURC and GOFRC work will continue in countries throughout the world.

Further, an international NSABB-like organization needs to be set up to manage a mutually agreed-upon approach for where and how DURC and GOFRC work should be done globally. This international organization should draw upon the guidance of experts in this area, not simply from the United States, but from around the world. We are under no illusions that such an approach would stop all intentional or unintentional misuse of newly emerging technologies. But to not try to stop it is irresponsible.

Priority 7: Recognize that TB, HIV/AIDS, malaria, and other life-threatening infectious diseases remain major global health problems.

The world can't afford to take its collective eye off of TB, HIV/AIDS, and malaria. In 2014, there were an estimated 36.9 million people living with HIV worldwide, resulting in 1.2 million deaths from AIDS. There were an estimated 9.6 million cases of tuberculosis, leading to 1.1 million deaths, according to 2015 statistics. And there were 214 million cases of malaria, with 438,000 deaths the same year. I fear the world hasn't fully come to grips with why it will become even more challenging to control, let alone dramatically reduce, the number of future TB and HIV/AIDS cases.

In 2014, it was estimated that only 63 percent of active TB cases were reported to the WHO, suggesting that more than 3 million infected and potentially infectious people were undiagnosed or unreported. The fact that TB control programs—often in HIV-infected populations—have been unable to get adequate funding, as well as the growing issue of antibiotic-resistant TB infections, does not bode well for global control. As we have painfully learned with the return of the *Aedes*-related diseases, public health gains from the past can quickly be lost if we let up on our efforts. The megacities of the developing world will only make the challenge of TB control more difficult.

The same forces are at play with HIV/AIDS, particularly in the developing world. A movement known as AIDS Free World looks to a day when there are effective vaccines and a cure for HIV. That is a wonderful aspiration, but if it inspires false hope that we are about to defeat HIV, that could cause a reduced sense of urgency among national governments and even possibly some philanthropic organizations to fund HIV/AIDS programs sufficiently.

Recent reports from countries in Asia—in particular, the Philippines—that new HIV infections are at an all-time high, as well as reports that the increasing number of new HIV cases in Africa outstrips treatment access provided by PEPFAR, speak to the enormity of the challenge. There is nothing in our public health playbook today that supports the UN target date of ending AIDS by 2030.

I feel more optimistic about the potential to control malaria because of the Gates Foundation's aggressive initiative, "Accelerate to Zero." Time will tell. But again, we must also remember the lessons of *Aedes,* playing out in Venezuela as we write this. In 1961, it was the first country in the world to be certified malaria-free. As a result of the collapse of the national economy, many thousands of financially desperate people migrated to the jungle mining areas in search of gold. The swampy mines where they work is a perfect breeding ground for malaria-transmitting *Anopheles* mosquitoes. Those who become ill with malaria return to their homes in the cities. There they spread the disease in squalid urban settings where there is no money for medicine or healthcare or spraying and mosquito control. In 2016, malaria has come roaring back. This is a vivid reminder that public health is intertwined with every aspect of life.

Priority 8: Anticipate climate-change effects.

As we detailed in chapter 4, climate change and a catastrophic pandemic are two of the four events that have the power to affect the entire planet. While climate change may not influence the likelihood of a pandemic, it surely will have a major impact on the incidence of other infectious diseases. Think of infectious diseases as fire and climate change as fuel. With climate change, some infections such as vector-borne diseases will put a substantially greater number of humans at potential risk as mosquito and tick populations grow in areas where they did not previously exist.

Climate change will also influence precipitation patterns, causing flooding and droughts, resulting in critical shortages of potable water and water used for crop irrigation. Rising sea levels will require the mass migration of densely packed groups of humans and animals from coastal lowlands, particularly in places like Bangladesh. Insufficient safe water and food will combine to create the perfect recipe for increasing the risk of infectious diseases.

We are only just beginning to understand the potential impact of climate change on infectious diseases in both humans and animals. We must maintain robust research and disease-surveillance programs to better understand and respond to this new normal.

Priority 9: Adopt a One Health approach to human and animal diseases throughout the world.

Throughout this book we have stressed the importance of the human-animal interface to the emergence and spread of infectious diseases. The time has come to address almost all human and animal infectious diseases as one continuum of risk and potential prevention and control. In the public health community this movement has become known as One Health. Today, we have the WHO for human health and the OIE, the World Organisation for Animal Health. The OIE's primary responsibility is to coordinate, support, and promote animal disease control. There are legitimate reasons from an animal health standpoint to have separate organizations—for example, some infectious diseases have major economic implications in food-production animals and not in humans. But until we recognize human and animal infectious diseases as one discipline, we will be disadvantaged in trying to prevent and control these diseases. We recommend that the WHO and OIE, as well as national government human health and animal health agencies, establish joint priority programs in One Health.

Now we come to the critical question of what kind of leadership, command, and control structure we need to make all this achievable—to be able to deal efficiently and effectively with the critical *who, what, when, where, why,* and *how* questions we enumerated at the beginning of this book.

One of the premises of our Crisis Agenda is that the United States will have to bear both the primary leadership responsibilities and the bulk of the financial burden. The G20 should provide substantial support, but given the relative lack of international support for global public health programs, this is unlikely to happen. Most of the G20 countries have provided only limited financial support for the WHO, have been largely absent in responding to critical regional outbreaks, and have undertaken minimal efforts in new vaccine and antimicrobial drug research and development.

The internal and external reviews of the WHO performance during the 2014–16 West African Ebola outbreak serve as important assessments of the capability of the international public health community and the WHO to respond to such a crisis. They should be considered seriously in discussions about reorganizing our global public health strategy. But the recommendations in these reports should be seen as just the beginning, not the complete agenda. For example, none of the reports addressed any of the highest-priority Crisis Agenda items we have identified.

We must clearly articulate what we need for global public health leadership and consider alternative approaches. Just as Lincoln had to go through a number of generals before he found one to lead the Union troops to victory, we may have to go through several iterations of an international public health infrastructure before we get it right.

In order to save ourselves as well as the rest of the world, we in the United States will have to step up. But the world, too, will have to realize a new level of public health leadership, organization, and accountability that will involve governments, the private sector, and philanthropic and nongovernmental organizations. It is one thing to say we need to commit *x* billions of dollars to fight our war against killer germs. But as anyone who has actually fought a war can tell you, all the resources in the world won't achieve much without leadership, accountability, and an effective command-and-control structure.

We strongly believe there must be a major overhaul of the WHO, beginning with its governance and financial support by member nations, for there to be any effective public health response to the twenty-first-century world of infectious diseases. If that cannot be accomplished, we need to start over and come up with a new international organization or agency that can do the job. The hallmark of such an agency would be its ability to strategically and tactically address the Crisis Agenda we've laid out. The US government must look carefully at both reprioritizing and reorganizing our own public health programs if we are to make meaningful change in how to prevent and control infectious diseases.

Laurie Garrett, the author of two important books, *The Coming Plague: Newly Emerging Diseases in a World Out of Balance* and *Betrayal of Trust: The Collapse of Global Public Health,* said to us, "I don't think that most of the people involved in global health today have adjusted to a twenty-first-century perspective on the problem sets and the solution sets. I think we're still looking at twentieth-century political realities, twentieth-century technologies, and twentieth-century perspectives on the scale of the problems. I think we'll be mired in paradigms that just as easily could have been taught in a school of public health in 1970 as in 2017."

The WHO is charged by the United Nations with promoting and protecting global health. But there are 194 member states, constituted as the World Health Assembly, and every single one of them gets an equal vote. As Bill Foege commented to us, "Imagine being the CEO of a corporation that had a 194-member board of directors!"

Despite the equal voting, most of the member nations provide little financial support, and authority is shared in a complex and uncomfortable tension between the director-general in Geneva and the regional headquarters around the world. With funding static for many years now and the virtual inability to get out ahead of an outbreak, it is no wonder the WHO was so roundly criticized for its response to the 2014–16 West African Ebola epidemic. Despite the lessons supposedly learned from the Ebola experience, the WHO has been criticized in 2016 both by African countries and by NGOs on the ground for its response to the yellow fever outbreak in Angola and the DRC.

Garrett expressed little optimism when she told us, "I've actually come to the point of feeling like it can't be reformed effectively. But we can make it a little better. We probably can't do without the WHO. But in the end, for the sorts of responses we really need, the sorts of capacities we desperately need to save lives worldwide, we need a completely, totally different 'think' about what we're up to."

Or as Bill Gates puts it, "WHO is not funded to do much. How many planes does it have; how many vaccine factories? We shouldn't think it's going to do things it was never intended to do."

Then there is accountability. The WHO is accountable to the World Health Assembly, which essentially means it is accountable to itself, or, no one.

Garrett notes, "All of the existing systems are without any concrete approach to accountability. There's no 'punishment.' There's no 'name and shame.' There's no price to be paid for failure or for screw-ups, for deliberate lying and cover-up. None of that will get you in serious trouble. If there's a court of adjudication of any kind, it's the court of public opinion. But the problem with the court of public opinion is that it was once fairly wide when it operated at the pace of newspapers. But the age of Twitter and Instagram has an attention span of ten seconds, so we don't have a mechanism where 'name and shame' results in a lasting reform."

If anyone outside the traditional scientific and political establishments has earned the right to be heard and listened to, it is Bill Gates, and more recently, Dr. Jeremy Farrar. The Bill & Melinda Gates Foundation and the US government together account for 23 percent of the WHO's budget, so that gives some idea of how influential the Gates Foundation is on the international public health stage. Jeremy Farrar has recently moved the Wellcome Trust into a similarly consequential global health role.

It is evident from even a brief conversation with Gates that he spends an enormous amount of time keeping up on the latest developments in the field, and not only in areas the foundation supports. Equally important, to reverse the old saying, he puts his mouth where his money is. Bill Gates has become a frequent and articulate commentator, analyst, and interpreter in the public health space in venues that range from TED Talks to the *New England Journal of Medicine.*

When we met with him, he offered a practical and sensible plan for using human and material resources already on the ground as the first assault wave against any emerging outbreak or epidemic.

> People aren't willing to pay for standby capacity [in public health]. They are in the military. They are in fire. I wish they would for epidemics, but they probably won't. And with standby capacity, you're never sure how good it is. We're starting this malaria-eradication effort, which is going to be region by region, and what I've decided is we should formalize the idea that as you're doing this disease eradication—let's just talk about malaria as an example—you have lots of good people on the ground. These guys know how to set up emergency operations centers, they know how to think about logistics, they know about messaging, they know about panic.

> We should say: Of these few thousand people, they are actually standby people for an epidemic. Because the malaria eradication is a super-important thing—I'm the biggest fan of it and will be very involved in it—but the nice thing is you can interrupt it.
>
> Worst case: You interrupt it for a year. Okay, the spread's back and it's bad. But it's got the people doing the kinds of things that you would need for an epidemic. So you can explicitly say, "Look, when we see problems, let's let thirty of those people look into it." "Okay, it looks real? Let's get all of them."
>
> That happened with polio [eradication efforts during the 2014–15 West African Ebola outbreak]. People don't acknowledge it and it wasn't formal. Nigeria is the place where you saw it most specifically. Yes, the Lagos [public health] people did a good job. But it was bolstered by the polio people [already working in the area] who came down and worked all throughout the system that had a major impact on Ebola.
>
> By tying these two functions together—the ongoing disease eradication programs and the emergency capacity—I think it will give visibility to both and maybe—net—get more resources.

As useful as this approach might be, it is not a substitute for an organization that can respond quickly and effectively to any infectious threat around the world.

Since the WHO can't fit this bill, who can?

In 2014, the US government launched the Global Health Security Agenda (GHSA) as a partnership among nations, international organizations, and nongovernmental stakeholders with the expressed aim "to help build countries' capacity to help create a world safe and secure from infectious disease threats and elevate global health security as a national and global priority." It now numbers fifty nations and is supposed to be supported by voluntary national assessments. A number of organizations, including the WHO, serve as advisers.

Like the WHO itself, I don't see how the GHSA can make a real difference in the Crisis Agenda. It may strengthen a country's healthcare delivery system and, potentially, its emergency response capability. But the GHSA has limited ability to impact diseases of pandemic potential or even of regional critical importance. Look no further than the public health emergencies of Zika and yellow fever: The GHSA has had little to no impact on the global response to these situations. It offers little leadership and support for global priorities such as vaccine research and development and the rapidly growing challenge of antimicrobial resistance.

Having spoken with numerous experts throughout the fields of public health and national and international governance, we believe a NATO-type treaty organization would be the best model to empower response to infectious disease crises. Member nations would precommit resources, personnel, and financial support so that the organization would be ready to react as soon as the threat becomes clear.

The most difficult part might be simply keeping politics out of it. "A treaty organization is good if you can get the kind of authority within it that's not going to be obstructionist," Tony Fauci comments. "I've got to tell you: That is really tough."

On the American domestic front, we have our own challenges in establishing effective public health governance and practice to meet the challenges of the twenty-first century. As a nation, we need to empower the leadership with resources and decision-making capacity as we do our military command structure, which makes decisions knowing their orders will be carried out and that resources needed to accomplish the mission are available. Just as important, the general officers know they are directly accountable for every decision they make.

Says Stewart Simonson, who served effectively under two HHS secretaries and had frequent interactions with the Oval Office, "There is a much more mature dialogue concerning national defense than there is for national preparedness."

Simonson cites the example of former governor Tom Ridge, when he was appointed by President George W. Bush to be the first secretary of Homeland Security after the 9/11 attacks. Ridge wanted to establish a functional operating model and set up regional commands, each headed by an officer—from FEMA, the coast guard, or a number of other agencies—who would be authorized to make decisions and move personnel, equipment, and funds to rapidly deal with an emergency.

Ridge's idea went nowhere, because no government agency wanted to subsume its own authority.

The most effective model for the kind of national entity we are talking about would likely require a governmental reorganization. We may now need a Department of Public Health, with its own cabinet secretary who can pull together the resources of the Department of Health and Human Services, including the Public Health Service, the NIH, the CDC, the FDA, and relevant parts of the Departments of Agriculture, Homeland Security, State, Defense, Interior, and Commerce. That office would have a much more focused set of responsibilities than the HHS secretary has today. For example, the Centers for Medicare and Medicaid Services, the organization within HHS that oversees nonmilitary healthcare services, had a fiscal year 2017 budget of approximately $1,012,765,000,000, while the combined CDC (infectious and noninfectious diseases) and National Institute of Allergy and Infectious Diseases (NIAID) of the NIH had a budget of $16,616,000,000. The CDC and NIAID budget is just 1.6 percent of the budget for Medicare and Medicaid, so it is easy to see where the HHS secretary has to direct a great deal of his or her attention. The new agency would also have a mandate and capacity for advance planning and quick global response, just like the Defense Department.

At a background briefing I gave members of the House of Representatives on Zika virus, one senior congressman commented that if we could show that each mosquito was actually a miniature drone controlled by ISIS, we could get all the funding we wanted.

Critical components of our military response are personnel, weapon systems, logistical support, intelligence, and diplomacy. We would not think of being without these resources or waiting to procure them until they are needed. If we have a crisis in the Mediterranean, we're prepared to send in a Sixth Fleet battle group. We don't start to requisition funds to build an aircraft carrier, two destroyers, a fleet of jet fighters, and everything else we would need.

To maintain the same level of preparedness in our ongoing war against infectious disease threats, we need to have personnel in place and ready to react: public health epidemiologists, physicians, nurses, veterinarians, sanitarians, statisticians, surveillance technicians, field-workers, lab personnel, and the support positions they all need.

Weapon systems include vaccines, antibiotics, pesticides, point-of-care laboratory tests, environmental health tools (wells, plumbing, and sewers), bed nets, and comprehensive global disease surveillance systems.

As far as leadership, I do not believe traditional public health professionals will be able to lead us out of our current infectious disease complacency. We need to have people who can see—and foresee—the big picture and know how to marshal the resources of government, science, and the private sector to face our challenges. These Crisis Agenda leaders need a unique understanding and practical expertise in global, regional, and national politics, as well as a critical working knowledge of the science behind the agenda. They need some of the same organizational talent that characterized Brigadier General Leslie Groves, the US Army Corps of Engineers officer who directed the Manhattan Project in World War II. They have to motivate governments and the public to support the Crisis Agenda, just as President Kennedy motivated the United States to get to the moon.

We know what we are suggesting will be difficult to implement and will require significant commitments of money, personnel, diplomacy, political power, and courage. That doesn't make it any less necessary. We shouldn't have to wait for something to happen before we react. The dots are there to be connected. When we say we were surprised by Zika, we shouldn't have been. When we say we were surprised by Ebola, or yellow fever, or chikungunya, or so many others, we shouldn't have been. And we shouldn't be surprised if tomorrow's crisis is caused by Mayaro virus, Nipah, Lassa, Rift Valley fever, or a new coronavirus.

And if, in the future, we are unprepared for a pandemic of a deadly strain of influenza, or antibiotics that no longer prevent common infections from causing serious or fatal illness, we certainly won't be able to say we weren't warned. Because we've had the warning and we have the solutions; we just need to act on them.

What can the average citizen do? Practically speaking, these are big, global problems that require big, global responses by powerful leaders and policy makers. But the average citizen can demand action. Our legislators, for instance, should never have been able to escape Capitol Hill in the summer of 2016 without passing bipartisan Zika funding. We've got to hold their feet to the fire and let them know in no uncertain terms that partisan politics has no place in public health policy or action. This will require the same kind of grassroots political action that it has taken to sway Congress on other issues.

CIDRAP advocates for the best science to implement proactive and nonpartisan public policy. I like to believe we are the citizens' representative on these issues. If you want to stay current and learn more about them, you can follow CIDRAP News and the other information on our website: www.cidrap.umn.edu. There is no charge, the information is updated daily, and you don't have to be a physician or scientist to understand it.

If we do start questioning and demanding as we should, and our leaders do start rising to their responsibilities in public health, will everything we've proposed and endorsed completely neutralize the threat of infectious diseases and the severe, even terrifying impact on modern life around the world? Of course not. But what we can do, with the necessary collective will and commitment of resources, is to give many more people throughout the world, particularly our children and grandchildren, the chance to live out normal, happy, and productive lives. And we can trade innumerable bad deaths for good ones.

And that is all we've ever hoped for.

Acknowledgments

From Mike: Along my personal and professional journey from boyhood home in Iowa to this book, I have been guided by many who gave of themselves tirelessly and selflessly. Words will never express my love and appreciation for so many who supported me and ignited my dream of a career in public health. That wouldn't have happened but for Les and Laverne Hull and Sarah Hill. In addition, Len Bruce, Tom Caulkins, David Duncklee, Ken Lampman, Ernie Lubahn, Marvin Strike, and the late Jim Wooden taught me to strive to make a difference.

Luther College was where my science and liberal arts education merged into the "right way" of seeing the world. Dr. Dave "Doc" Roslien took charge of that process and is still there for me, as are his enlightened colleagues Wendy Stevens and Drs. Jim Eckblad, Roger Knutson, Phil Reitan, John Tjostem, and the late Russ Rulon, and their spouses.

Luther handed me off to the University of Minnesota School of Public Health (SPH) in 1975, and I have never left. Even during twenty-four years at the Minnesota Department of Health, my academic home was SPH. Soon after my arrival, the late professor Rex Singer became my academic touchstone, personal adviser, and dear friend. His was an immeasurable legacy for me and the hundreds of others he shepherded.

As I had at Luther, I benefited immensely from a team of best and brightest senior academicians who invested in me and my work well beyond anything I deserved. They include the late dean Lee Stauffer, deans Drs. Mark Becker and John Finnegan, and the late Drs. R. K. Anderson, Velvl Greene, Leonard Schuman, and Conrad Straub. More recently, that support has continued from Drs. Frank Cerra, Aaron Friedman, Brooks Jackson, and Tucker LeBien.

Dr. Barry Levy hired me at the Minnesota Department of Health, taking a flier on a kid who had just started graduate school. My group and I went from inexperienced young epidemiologists to a finely oiled team that could take on the most difficult infectious disease mysteries. There I had the opportunity to mentor two individuals who would eventually surpass their teacher. Drs. Kristine Moore and Craig Hedberg, both now at the University of Minnesota, are at the top of my list of valued professional and personal relationships. Kris is CIDRAP's medical director, and Craig is a professor in the Division of Environmental Health Sciences. Others at the Minnesota Department of Health who made major contributions to my career are Sister Mary Madonna Ashton, Kristen Ehresmann, Jan Forfang, Linda Gabriel, Ellen Green, the late Jack Korlath, Aggie Leitheiser, Lynne Mercedes, Michael Moen, Terry O'Brien, Joan Rambeck, Mary Sheehan, John Washburn, Karen White, Jan Wiehle, and Drs. Jeffrey Bender, John Besser, Richard Danila, Kathy Harriman, Ruth Lynfield, and Kirk Smith.

Today my professional home is CIDRAP. Its creation was possible only because of Michael Ciresi and Kathryn Roberts. In addition to Kris, Jill DeBoer and Elaine Collison, also health department veterans, make up the leadership team. My respect and admiration for them are unlimited. Marty Heiberg Swain is a founding CIDRAP member and one of the best editors on the planet. Julie Ostrowsky, Lisa Schnirring, and Jim Wappes are also invaluable colleagues. Former CIDRAP employees Aaron Desmond, Karina Milosovich, and Robert Roos gave of themselves generously to make us what we are today. Nicholas Kelley, my former PhD student and CIDRAP employee, taught me as much as I ever taught him. Over the past fifteen years Judy Mandy and Laurel O'Neill have run the day-to-day operations. They are my air traffic controllers and touchstones to reality.

CIDRAP has been able to pursue its efforts because of the generous support from donors who understand the importance of our mission, in particular the Bentson Foundation and the unwavering contributions of Laurie Bentson and Judi Dutcher.

Following the events of 9/11, HHS secretary Tommy Thompson asked me to split my time between the University of Minnesota and serving as his special adviser. I performed both roles for almost three years, during which time I came to know him as a dynamic, visionary, and caring leader and close friend. I was also privileged to work closely with his successor, Michael Leavitt, another leader I hold in the highest personal and professional regard. Stewart Simonson served both secretaries in critical roles, and there have been few senior US government officials more capable, humble, and accomplished.

I have been blessed by priceless tutoring and unwavering support from some of the giants in my field: the late William Patrick and the late Drs. William Hausler, Edward Kass, Joshua Lederberg, William Reeves, Sheldon Wolff, and John "Jack" Woodall, as well as Drs. William Foege, Philip Russell, and Alfred Sommer.

Thank you, special friends and most respected colleagues: Drs. Massoud Amin, Edward Belongia, Ruth Berkelman, Seth Berkley, Robert Bowman, Becky Carpenter, Gail Cassell, James Curran, Jeffrey Davis, Martin Favero, David Franz, Bruce Gellin, Richard Goodman, Dan Granoff, Duane Gubler, Margaret Hamburg, Penny Heaton, Thomas Hennessy, Keith Henry, James Hughes, David Ingbar, Allan Kind, Amy Kircher, Joel Kuritsky, Jody Lanard, Monique Mansoura, Thomas Monath, Trudy Murphy, James Neaton, Gerald Parker, Phillip Peterson, George Poste, David Relman, Peter Sandman, Patrick Schlievert, James Todd, Pritish Tosh, and David Williams. In addition: John Barry,

Richard Danzig, Susan Ehrlich, Larry Gostin, Diana Harvey, Ann Leon, Gina Pugliese, Don Shelby, Janet Shoemaker, Kristin Stouffer, and Sarah Youngerman.

Two individuals deserve my deepest and very special gratitude, appreciation, and love for their support of my professional and personal life. Drs. Julie Gerberding and Walter Wilson are highly respected colleagues who also are in every way my adopted sister and brother.

My associates at the National Institutes of Health have supported our work in so many ways. Dr. Anthony Fauci is a critical figure in our business, but I most value our special friendship over these past thirty-plus years. Others at the NIH include the late Dr. John LaMontagne and Drs. Carole Heilman, Linda Lambert, Pam McInnes, and Diane Poste. Thanks also to Greg Folkers.

John Schwartz coauthored *Living Terrors*. To this day I appreciate the lessons he shared with me as a talented writer and friend.

Last but not least, this book is all about and because of my family. I can only imagine what the world will be like for you, my kids and grandkids, if our battle with infectious diseases doesn't change course. If there is anything I can do to alter that course, my entire career will be worth the effort.

Libby, thank you for being the sunshine of my life. Your understanding and encouragement of my passion for the world of public health is a gift. Your love is priceless.

From Mark: I always count on the knowledge, experience, and counsel of my two medical brothers, Drs. Robert and Jonathan Olshaker, and Robert's wife, Dr. Jacqueline Laurin. Their careers and care for their patients are a living tribute to my late father, Dr. Bennett Olshaker.

For more than three decades, I have benefited from the collaboration with my film-producing partner Larry Klein—among the top science producer-directors in the business—who has been perceptive enough to use Mike repeatedly as a program participant and adviser. Larry's influence is reflected throughout this book.

Marty Bell—distinguished writer, Broadway producer, and now political advocate—got me into book writing and has been a constant source of encouragement, support, and ideas. Every writer should have a posse of literary compadres. Mine is fortunate to include: Jeff Deaver, Eric Dezenhall, John Gilstrap, Jim Grady, Larry Leamer, Dan Moldea, Peter Ross Range, Jim Reston, Gus Russo, Mark Stein, James Swanson, Joel Swerdlow, and Greg Vistica.

My wife, Carolyn, has been not only my partner in all things, but an enthusiastic fellow traveler through all the adventures, as well as attorney, manager, counselor, and inspirer. I love you more than anything and couldn't have done it without you.

This book has been a true collaboration, but not just between the two of us.

At the top of our team is our editor, Tracy Behar, who had faith in us and vision to see the book for what it could be. Her nurturing, advice, gentle prodding, and meticulous editing steered us through, shaping our narrative and honing our message. Every writer should be lucky enough to have an editor and friend like Tracy. Happily for Little, Brown, these traits are shared by Senior Vice President and Publisher Reagan Arthur, who also believed in us from the beginning.

Our agent, Frank Weimann of Folio Literary Management, was immediately enthusiastic about the project, guided us through proposal and presentation, and encouraged us through every step of writing.

In addition to those already mentioned, we tremendously appreciate the sizable contributions to this book of Dr. Barry Beaty, Dr. Martin Blaser, Dr. James Curran, Dame Dr. Sally Davies,

Laurie Garrett, Bill Gates, Dr. Dwayne Gubler, Ron Klain, Maryn McKenna, Lord Dr. Jim O'Neill, Stewart Simonson, Dr. Brad Spellberg, and Dr. Lawrence Summers. Thanks to Julie Clemente for research and updates in our wide-ranging fields of concern.

We take this opportunity to remember with love and gratitude our late attorney and friend Steven Paul Mark, who encouraged us from the beginning, got us together with Frank, and was so vital in putting this project together. He is very much missed.

Finally, there is Dr. Donald Ainslie "D.A." Henderson, who passed away shortly after we completed this book. D.A. was a tough, courageous, true hero of the world. Through his leadership of the campaign to eradicate smallpox, he is probably responsible for averting more early deaths than any individual in history. In his illustrious career, D.A. was a public health visionary, inspiring mentor, powerful moral presence, and very dear, special friend. Through the example of his life, he showed us all what is possible.

Index

acquired immunodeficiency
syndrome. *See* HIV/AIDS
Adadevoh, Ameyo, 154
Adleman, Leonard, 134
AIDS Free World, 307
See also HIV/AIDS
Alibek, Ken (Kanatjan Alibekov),
128–29
Alibek, Lena, 129
American College of Obstetricians
and Gynecologists, 40–41
Amherst, Jeffery, 128
AMR Centre, 246–47
AMR (antimicrobial resistance)
studies, 218–19, 227, 230, 240,
249–50, 251–52
amyl nitrite, 12
animal reservoirs, 3–4, 64–65,
78–79, 112, 142, 309
for Ebola, 73, 144, 145, 150
for influenza, 68, 117, 160, 256–57,
264–67, 286–87
for La Crosse encephalitis, 183
for malaria, 182
for MERS, 168–71
for SARS, 164, 166, 167
for Western equine encephalitis, 186
for Zika, 205
animals
antibiotic use in, 221, 228–36,
245, 251
population of, 68
research using, 115, 117, 119
anthrax, 65
antibiotics for, 131
biowarfare using, 70–71, 92, 114,
126–27, 128, 129, 130–32, 143
transmission of, 77
vaccine for, 94–95
antibiotics, 26, 216–17
animal use of, 221, 228–36,
245, 251
for anthrax, 131
discovering new, 219–20, 246–48
human use of, 222–28
labeling of, 243–44, 248
living without, 220–21
overuse of, 222, 223, 243, 274–75
preventing need for, 238–41
protecting efficacy of, 241–46, 251
for sexually transmitted
diseases, 75
for TB, 99, 217
true value of, 248
antigenic drift or shift, 255, 286, 293
antimicrobial resistance, 215–16
animal antibiotic use and, 228–36
antibiotic efficacy and, 241–46
awareness of, 250–53, 302
categories of, 221–22
global threat of, 55, 56, 300
human antibiotic use and, 222–28
infection prevention and, 238–41
novel solutions for, 248–50
policy and plan for, 237–38
priorities for, 238–43
reports and studies on, 218–19
surveillance system for, 240–41
TB and, 108, 110, 228, 307
trends in, 219–21
antiviral drugs
for HIV, 17, 20, 75, 99, 104, 107
for influenza, 256, 273, 275, 276
for MERS, 172

Apley, Michael, 230
Ariel, Will, 24
artemisinin, 100
arthropods, 73, 181
aureomycin, 229
autism, vaccines and, 84, 288
avian influenza, 256–57
  human infection with, 283–84
  pandemic potential of, 262, 264–67
  reservoirs for, 68, 79, 160
  scenario for, 271–83
  viruses for, 116–20, 123, 260, 262, 265–67

bacteria
  evolution involving, 59
  gram-positive and -negative, 225
  gut, 60, 223
  novel strategies against, 249
  See also antimicrobial resistance
bacteriophages, 249
Barry, John, 256, 259
bats, 73, 144, 145, 164, 167, 168, 169
Behring, Emil von, 249
Berkley, Seth, 80, 156–57
Bhullar, Kirandeep, 215
Biden, Joe, 156
Bill & Melinda Gates Foundation, 99–100
  funding by, 89, 110, 313
  malaria strategy of, 100, 102–4, 304–5, 307–8
  See also Gates, Bill
biomarker tests, 250
Biomedical Advanced Research and Development Authority (BARDA), 94–95, 142, 246–47
biosecurity, 229
bioterrorism, 125–26, 300
  agents for, 128–30, 134–35
  anthrax in, 70–71, 77, 92, 114, 126–27, 128, 129, 130–32, 143
  bipartisan report on, 305–6
  countermeasures for, 92–94, 131–32
  history of, 127–28
  preparedness for, 139–43
  prevention of, 55–56
  smallpox in, 128, 129, 130, 132–34, 135, 136, 137–39
bird flu. *See* avian influenza
birds, 117, 160, 182, 186, 256–57, 264–65
Black Death, 65–66, 128, 259, 283
Black Swan, The (Taleb), 13
Blaser, Martin, 59–60, 218, 230, 232, 233
blood transfusion, 14–15, 17–18, 210
Blue Ribbon Study Panel on Biodefense, 305–6
Brainerd diarrhea, 41–47
Buffalo Springfield, 7
Bush, George W., 106, 316
Butler, Richard, 125–26

camels, 168–71
Cameron, David, 218
cancer, 217, 219
candidal infection, 10–11
carbapenems, 234
Carson, Rachel, 193
Center for Infectious Disease Research and Policy (CIDRAP), 4, 56, 318–19
  antimicrobial stewardship site of, 246
  drug survey by, 263–64
  influenza vaccine reports by, 121, 176, 287–88, 290, 293–94
  tabletop exercises of, 270
Centers for Disease Control and Prevention (CDC), 122–23, 316
  antibiotics and, 223, 240, 245
  Brainerd diarrhea and, 44–47
  Ebola and, 154
  HIV/AIDS and, 7–15, 19
  influenza and, 115, 288, 289
  mosquito-borne diseases and, 186, 188
  toxic shock syndrome and, 30–41
  Zika and, 208
Centers for Medicare and Medicaid Services, 239, 316
Chain, Ernst, 216, 217, 220
Chan, Margaret, 146, 205
Chen, Johnny, 159, 160, 163, 165
chicken pox, 65, 83–84
chickens, 68, 233, 235, 273
chikungunya, 78, 201–2
  genetic changes in, 212
  vector for, 101, 189, 200

chimera agent, 112–13
Chimera Project, 129
cholera, 27–28
Christie, Chris, 148
Churchill, Winston, 21, 158, 286
CIDRAP. *See* Center for Infectious Disease Research and Policy
ciprofloxacin (Cipro), 225, 231, 235
civets, 164, 166, 167
Clapper, James R., 114
climate change, 52, 70, 269, 300
  anticipating effects of, 308–9
  belief in, 25
  international effort on, 244, 302
  vector-borne diseases and, 78, 182
Coalition for Epidemic Preparedness Innovations (CEPI), 97, 158, 291, 302–4
colistin, 234–35
Collins, Francis, 119
Compelling Need for Game-Changing Influenza Vaccines (CIDRAP), 121, 290–91, 293–94, 301
contraception, access to, 105
coronaviruses, 55, 164–65, 167–68
cowpox, 26, 65, 80
Crick, Francis, 133
crime, disease compared to, 5–6, 57–58
Crisis Agenda, 299–301
  leadership for, 309–11, 317–18
  priorities of, 55–56, 300, 301–9
CRISPR (clustered regularly interspaced short palindromic repeats), 113–14
Crozier, Ian, 152
Cunnion, Stephen, 161
Cuomo, Andrew, 148
Curran, James, 8–10, 12, 13–15, 16–17, 74
Cushing, Harvey, 30
cytokines, 63, 258
cytomegalovirus, 10–11, 210
Czaplewski, Lloyd, 247

Dalai Lama, 178
Danzig, Richard, 306
Darrow, Bill, 9
Darwin, Charles, 112
Daschle, Tom, 126–27
Daszak, Peter, 167
Davis, Jeffrey, 31–32, 33, 35
DDT, 189, 193, 202
death, bad *vs.* good, 25
dengue, 5, 197–201, 300
  genetic changes in, 212
  global spread of, 78, 213
  serotypes of, 198–99
  sexual transmission of, 210
  vaccine for, 195, 199–200
  vector for, 101, 188, 189–90, 200–201, 214
dengue hemorrhagic fever (DHF), 197, 198, 199–200
diagnostic tests, 250, 303–4
diarrheal diseases, 56, 300
  antibiotic-induced, 223
  antibiotics for, 227
  Brainerd, 41–47
  waterborne, 240
Dickens, Charles, 68, 299
diphtheria, 82, 83, 85, 249
disaster management, 50–53, 269
DNA (deoxyribonucleic acid), 62, 113–14, 133
Doctors Without Borders, 156, 159
Domagk, Gerhard, 216–17, 220
Douglas, John, 136
drugs
  generic, 237, 264
  guidelines for using, 243–44, 248
  production of, 69, 86
  revenue from, 85
  stewardship of, 241
  stockpiles of critical, 263–64, 275
  See also antibiotics; antiviral drugs
dual-use research of concern (DURC), 111–12, 300
  example of, 117–20, 122–23
  issues for, 55–56, 120–21, 124
  minimizing use of, 306
  pros and cons of, 114
Durant, Will, 24

Eastern equine encephalitis, 182
Ebola, 144–58
  2014 outbreak of, 54, 70, 71, 145–50, 153–56, 310, 312
  biowarfare using, 128

Ebola *(cont.)*
burial practices for, 26
containment of, 212
diagnostic tests for, 304
international response to, 284–85
latent cases of, 152–53
pandemic potential of, 53, 109
prevention of, 55
research on, 123
reservoir for, 73, 144, 145, 150
Reston, 65, 150
symptoms of, 146–47
transmission of, 79, 145, 146, 147–52, 259
vaccine for, 89, 90, 95–96, 156–58, 177, 303
virus causing, 144, 146
Zaire, 65, 150–51
eggs, vaccines grown in, 86, 196, 287, 294
Eisenberg, Barry, 241
Eisenhower, Dwight D., 51
electricity, advent of, 29
Enbrel, 85
encephalitis, 129, 179–80, 182–88
Englund, Bruce, 163
enterotoxin type B, 31, 40
Environmental Protection Agency, 143
Epidemic Intelligence Service (EIS), 9, 30
epidemics. *See* pandemics
epidemiology, 22, 48
consequential, 3
goals and tools of, 26
observation method in, 26–29
purview of, 47
questions in, 7–8
shoe symbol for, 30
threat matrix in, 49
Epstein-Barr virus, 11
Escherichia coli, colistin-resistant, 235
Eumenes II, King of Pergamum, 128
European Centre for Disease Prevention and Control (ECDC), 207
European Organization for Nuclear Research (CERN), 247
evolution, microbial, 58–59, 60–61, 112–13
of Ebola virus, 146, 151
of influenza virus, 121, 264–65

Farrar, Jeremy, 158, 313
Fauci, Anthony "Tony," 93, 107–8, 288, 293, 294, 315
FDA (US Food and Drug Administration)
antibiotic regulations of, 230, 232–33, 244
tampon safety and, 35
vaccine approval by, 86, 87–88
ferret-badgers, 164, 166, 167
ferrets, 115, 117, 119
Fink, Gerald, 114
Fink Report, 114–15
Finland, Max, 219
Finlay, Carlos, 190, 191, 192
flavivirus, 190, 210
Fleming, Alexander, 215, 216, 217, 220
Florey, Howard, 216, 217, 220
flu. *See* influenza
fluoroquinolones, 231
Foege, William "Bill," 3, 22–24, 48, 64, 138, 311
fomite, 192
Ford, Gerald, 260
Fouchier, Ron, 116
Foundation for Vaccine Research, 95, 96–97
Frankenstein (Shelley), 111
Franz, David, 125–26
Frieden, Tom, 228, 235

Gallo, Robert, 15
Garrett, Laurie, 311, 312
Gates, Bill, 23, 99–100
on bioterrorism, 141
on epidemic risk, 50, 254
on resource utilization, 313–14
vaccine development and, 89, 90
on World Health Organization, 312
See also Bill & Melinda Gates Foundation
Gates, Melinda, 23, 99–100
genetic engineering, 112–14, 203
See also GOFRC (gain-of-function research of concern)

genetic mutation, 62, 63
  antibiotic resistance and, 217, 218, 234–35
  beneficial, 217
  in influenza virus, 68, 76–77, 116, 255, 267, 286, 293
  intentionally created, 111–12
  in MERS virus, 169, 171
  in Zika virus, 152, 212–13
genetic reassortment, 255, 256, 257, 273
germ theory, 191
Gibson, William, 144
Global Alliance for Control of *Aedes*-Transmitted Diseases (GAAD), 204, 304–5
Global Health Initiative, 106
Global Health Security Agenda (GHSA), 314–15
Global Leadership Against HIV/AIDS, Tuberculosis, and Malaria Act, 103
GOFRC (gain-of-function research of concern), 111–12, 300
  assessing and funding, 123–24
  biological weapons and, 129–30, 133–34
  CRISPR tool for, 113–14
  issues for, 55–56, 120–21, 124
  minimizing use of, 306
Gore, Al, 156
Gorgas, William C., 190, 192
Goyan, Jere, 35
Gram, Hans Christian, 225
Gretzky, Wayne, ix
Groves, Leslie, 318
growth promotion, antibiotic use for, 229–30, 232, 245
Gubler, Duane J., 193, 194, 196, 200, 203, 204, 212, 214
Guillain-Barré syndrome
  influenza vaccine and, 260
  Zika and, 5, 48, 206–7, 208–9, 211
Guinan, Mary, 9

Haldane, J. B. S., 58
Halstead, Scott, 199–200
handwashing, 239
Hannibal, 128
Hardin, Garrett, 222
Harness, Ed, 36–37
Hatchett, Richard, 212
Hausler, William, 185–86
heart disease, 217, 219
Heckler, Margaret, 15, 16, 18
hemagglutinin (HA), 76, 255–56, 286, 287, 293
hemophilia, 14–15
Henderson, D. A., 3, 93
hepatitis, 8–9, 10, 62, 107, 300
herpes zoster, 65
Hewitt, Don, 126
Heymann, David, 161–62
HIV/AIDS, 8–20, 74, 104–8
  biowarfare using, 128
  case surveillance for, 12–14
  differential diagnosis of, 11–12
  diseases associated with, 8–11, 13–14
  drugs for, 17, 20, 75, 99, 104, 107
  global threat of, 52, 300, 306–8
  prevention of, 56, 105
  retrovirus causing, 15, 16, 134
  screening for, 19
  statistics on, 19–20, 98, 104, 107, 306–7
  TB with, 108–9, 307
  transmission of, 14–15, 16, 17–18, 77–78, 147
  vaccine for, 16–17, 18–19, 89, 104, 107–8, 288, 296–97, 301
honey bees, 187, 188
horses, 186
hospitals, infection control in, 26, 165, 166–67, 173, 239
host, disease, 61–62, 249
Hot Zone, The (Preston), 65, 150
Hull, Laverne, 22
Human Genome Project, 133
human immunodeficiency virus. *See* HIV/AIDS
Humira, 85
hurricanes, 51

immune enhancement disease, 198, 199
immune response
  cytokine storm in, 63, 258
  suppression of, 10–11
  to vaccines, 16, 62–63, 294

infectious diseases
antibiotic prophylaxis for, 229
antibiotic-resistant, 220, 238–41, 248–49
black swan of, 13, 20
crisis and major, 3–4, 54–55, 99, 300
diagnostic tests for, 250, 303–4
endemic, 56, 70–71, 300
health care facilities and, 26, 165, 166–67, 173, 238–39
life expectancy and, 29, 217, 259, 261
metaphors for, 5–6, 24–25
One Health approach to, 309
outbreaks of, 5, 33, 64–70
pandemic and epidemic, 25, 52–53, 55, 71, 300
passive treatment of, 249
preparedness for, 50–51, 95–97, 317–18
prevention of, 26
of regional importance, 53–54, 55
reproductive rate for, 165–66
resource utilization for, 313–14
superspreaders of, 165–66, 167
surveillance of, 13–14, 240–41, 250
transmission of. *See* transmission
influenza, 254–67
drugs for, 256, 273, 275, 276
reservoirs for, 68, 117, 160, 256–57, 264–67, 286–87
SARS and, 160, 161
surveillance system for, 240–41
transmission of, 73–74, 76–77, 117, 119, 148, 267, 272
See also avian influenza
influenza pandemics, 268–85
1918, 63, 66, 73–74, 115, 121, 257–59, 261–62, 268, 269–70
1957/1968, 121, 259–60, 261–62
2009, 87, 90, 115, 256, 258, 260–62, 289
case patterns in, 261–62
consequences of, 262–64, 297–98
history of, 257–61
potential for, 52, 55, 56, 62, 117–18, 119, 257
preparedness for, 268–70, 291–92
scenario for, 270–83, 291
influenza vaccines, 84, 85
annual formulation of, 91, 254–55, 286
campaign for, 260
CIDRAP report on, 121, 290–91, 293–94, 301
efficacy of, 90–91, 287–90
game-changing, 91, 176, 282, 292–97, 301
production of, 86–87, 196, 287, 288–89, 295–96
Russian studies on, 122
scenario for, 273, 276, 277–78, 279–80
influenza viruses, 254
antigens of, 76, 255–56, 286, 287, 293
H1N1, 87, 90, 115–16, 121–22, 256, 258, 260–61, 289
H2N2 and H3N2, 121, 260, 261, 288
H5N1, 116–20, 123, 260, 262, 265
H5N2 and H5N6, 266–67
H7N9, 262, 265–66, 271–83, 283–84
hyperevolution of, 264–65
identification of, 286–87
mutation of, 68, 76–77, 116, 255, 267, 286, 293
Institute of Medicine, 41, 96, 172, 269
insurance industry, 297–98
Intergovernmental Panel on Climate Change (IPCC), 302
International AIDS Vaccine Initiative (IAVI), 296–97
Ivins, Bruce, 127

Jacobson v Massachusetts, 81
Jefferson, Thomas, 81, 83
Jenner, Edward, 26–27, 80–81, 192
Jobs, Steve, 113
Jukes, Thomas, 229
Juranek, Dennis, 9

Kaczynski, Theodore, 136, 163
Kaposi's sarcoma, 8, 11, 12–13, 17–18
Kawaoka, Yoshihiro, 116
Kennedy, John F., 318
Kim, Jim Yong, 98

Kipling, Rudyard, 159, 161
Klain, Ron, 155–56, 284–85
Kwan, Sui-chu, 163

La Crosse encephalitis, 179–80, 182–86
LaMontagne, John, 93
Lantus, 85
Laxminarayan, Ramanan, 231, 245, 246
Leahy, Patrick, 127
Leavitt, Michael, 115, 291
Lederberg, Joshua, 24, 237
Lemierre's syndrome, 226
Leonardo da Vinci, 72
Levi, Primo, 23
Lieberman, Joseph, 139–40
life expectancy, 29, 217, 259, 261
Liu Jianlun, 162–63, 165
Living Terrors (Osterholm), 92, 125
lysins, 249

MacArthur, Douglas, 49
MacGillivray, Greg, 51
Madison, James, 81
Mahoney, Frank, 154
malaria, 5, 100–104
  climate change and, 70
  drug-resistant, 78
  eradication of, 313–14
  Gates' strategy for, 100, 102–4, 304–5, 307–8
  global threat of, 99, 300, 306–8
  prevention of, 56
  statistics on, 98, 101
  vaccine for, 89, 102
  vector for, 182
Mall of America (Minnesota), 132
Manhattan Project, 296, 301, 318
Marburg filovirus, 73, 144
March of Dimes, 89
McKenna, Maryn, 228, 233–34, 245, 248
McVeigh, Timothy, 135
measles, 66
  German, 210
  reemergence of, 84–85
  transmission of, 65, 76, 165
  vaccine for, 82, 83, 288
medical countermeasures (MCM), 142–43
Mendel, Gregor, 112
Mendelson, Sarah, 106
meningitis, bacterial, 179
MERS (Middle East respiratory syndrome), 167–76, 300
  preparedness for, 174–76
  prevention of, 55
  regional outbreaks of, 4, 54, 167–74
  reservoir for, 168–71
  superspreaders of, 168, 172–73
  vaccine for, 95, 172, 176, 303
Meselson, Matthew, 125–26
miasma theory, 27–28, 191
microbes, 57–63
  classification of, 61
  evolution of, 58–59, 60–61, 112–13
  human relationship with, 57–58, 59–60
  immune response to, 62–63
  mutation in, 62, 63, 217
  preventing harmful release of, 55–56
  reproduction by, 60
  See also bacteria; pathogens; viruses
microbiome
  global, 58, 60
  personal, 60, 218
microcephaly, Zika and, 48, 207–10
military response, components of, 317
Minnesota Department of Health, 10, 42–47
Missing Microbes (Blaser), 60, 218
Mitchell, John Kearsley, 191
monkeypox, 65, 136–37
monkeys, 150, 182, 205
mononucleosis, infectious, 11
Montagnier, Luc, 15
Moore, Kristine (née MacDonald), 44–45, 47, 179
Morbidity and Mortality Weekly Report (*MMWR;* CDC), 10–11, 14, 15, 32, 34

mosquitoes, 180–81
Aedes, 101, 191, 193, 200, 201, 202, 203, 209, 214, 300, 304–5
Aedes aegypti, 79, 188–90, 191, 193–94, 200, 202, 205, 207, 304
Aedes albopictus, 188, 202, 207
Aedes triseriatus, 180, 182–86, 189
Anopheles, 101, 102, 305, 308
control of, 102–3, 104, 187–90, 192–93, 200–201, 202–3, 214, 304–5
Culex tarsalis, 186
disease transmission by, 5, 55, 70, 73, 78, 101, 178, 181–82, 189, 200, 204, 300
La Crosse encephalitis and, 179–80, 182–86
Western equine encephalitis and, 182, 186–88
See also chikungunya; dengue; malaria; yellow fever; Zika
Motsoaledi, Aaron, 109
MRSA (methicillin-resistant *S. aureus*), 220
Müller, Paul Hermann, 193
mumps vaccine, 82, 83
Murrow, Edward R., 83

Nagarajan, Ganesh, 235
naled, 214
Narayan, Adi, 235
National Academy of Medicine, 41, 96, 172, 269
National Blueprint for Biodefense (Lieberman & Ridge), 139–41, 142–43
National Institute of Allergy and Infectious Diseases (NIAID), 102, 142–43, 316
National Science Advisory Board for Biosecurity (NSABB), 115–20, 123–24, 306
Nelmes, Sarah, 26
neuraminidase (NA), 76, 255–56, 286, 293
Nipah virus, 140–41, 300
Nixon, Richard, 128
Nobel, Alfred, 159
Norwegian Institute of Public Health, 95, 96–97

Obama, Barack, 106, 155–56, 218, 229
observation method, 26–29
Olshaker, Jonathan, 226
Olshaker, Mark, 51, 89, 102, 128–29, 131, 136, 226
One Health, 64–65, 142, 172, 309
O'Neill, Jim, 218–19, 252
Oparin, Alexander, 58
organ tropism, 62
oseltamivir (Tamiflu), 256, 273, 275
Osler, William, ix, 220
Osterholm, Erin, 34, 180
Osterholm, Michael T., 3, 4
60 Minutes interview with, 125–26
CIDRAP founding by, 4, 56
family crises of, 17–18, 21–22, 178–80
heroes of, 22, 24, 27, 28
Living Terrors, 92, 125
mosquito-borne disease and, 178–80, 182–86, 201
on next pandemic, 6
Osterholm, Ryan, 178–80, 182
Ottawa Indians, 128

Pacini, Filippo, 27
Pan American Health Organization, 189, 193
pandemics, 52–53
impact of, 65–66, 69, 71, 308
pathogens causing, 55, 56, 117–18, 300
preparation for, 6, 53
threat of, 25, 50, 254, 297–98
See also influenza pandemics
Park Geun-hye, 173
Pasteur, Louis, 45, 81
pasteurization, 45–46
pathogens, 58, 95
critical regional, 53–54, 55, 300
pandemic, 55, 56, 117–18, 300
threat matrix for, 49
transmission of, 72–79
See also bacteria; viruses
Patrick, William "Bill," 131
Pattison, Kermit, 3
Pauling, Linus, 126
Peace Corps, 71
Pearson, Natalie Obiko, 235

penicillin, 215, 216, 217, 220
pentamidine, 9, 14
permethrin, 103
pertussis vaccine, 81–82, 83, 85
pesticides
developing new, 102, 304, 305
for mosquito control, 103, 104, 187–88, 189, 193, 202–3, 214
pharmaceutical companies, 69
antibiotic development by, 220, 246–48
on drug-resistant infections, 252–53
labeling guidelines for, 243–44, 248
vaccine development by, 83, 84, 85–96, 157–58, 176, 196, 199, 220, 295, 302–4
Phipps, James, 26
pigs, 68, 256, 257, 264, 266–67, 286–87
piperacillin-tazobactam (Zosyn), 224–25, 244
plague, 65–66, 128, 259, 283
plasmids, 234
plasmodium, 101
pneumococcal vaccine, 240
pneumonia
bacterial, 300
influenza and, 258–59
Pneumocystis carinii, 8, 9, 10, 11, 12–13, 14, 18
Poe, Edgar Allan, 268
polio, 65
eradication of, 313–14
Gates' war on, 100
Nigerian program for, 154
transmission of, 165
vaccines for, 82–83, 89
virus for, 133–34
population
concentration of, 64, 65, 154–55
of First World countries, 237
growth in, 16, 67–68, 78
pregnancy, Zika infection in, 207–12
President's Emergency Plan for AIDS Relief (PEPFAR), 106–7, 307
President's Malaria Initiative (PMI), 103–4
Preston, Richard, 65, 150
primordial soup, 58
Project Bioshield, 92–93, 94
prontosil, 216–17
Pseudomonas aeruginosa, 224–25
public accounting and commitment, 242–43
public education, 26, 244–45
public health, 3–4, 22–23
agenda of, 25
disaster management in, 50–51
domestic strategy for, 315–19
father of modern, 29
federal funding for, 107
Foege's tenets of, 23–24
global strategy for, 67, 309–15
leadership in, 309–11, 317–18
patron saint of, 27
private partnerships with, 91–97, 246–48, 296–97, 301, 302–4
proactive practice in, 188
threat matrix in, 49–50

quarantine, 26, 139, 148–49, 154, 173
quinine, 100

rabies, 61–62, 81, 84, 99
Racaniello, Vincent, 151–52
Raub, William, 93
Reed, Walter, 190, 191
Reeves, William "Bill," 188
Regan, Thomas F., 17, 18
Relenza (zanamivir), 256, 273, 275, 276
Remicade, 85
retrovirus, HIV, 15, 16, 134
Ridge, Thomas, 139–40, 316
Rising Plague (Spellberg), 222
RNA (ribonucleic acid), 62, 255
Rockefeller Foundation, 189, 193
rodents, 183
Roosevelt, Franklin D., 276
Rosen, Otto Karl von, 128
Roueché, Berton, 22
rubella, 82, 83, 210
Rubio, Marco, 212
Rumsfeld, Donald, 269
Rush, Benjamin, 198
Russell, Philip K., 93, 125
Ryan, Romana Marie, 17–18

Sabin, Albert, 83
Salk, Jonas, 57, 82–83
Samsung Medical Center, Seoul, South Korea, 173–74, 175
sanitation, 26, 82, 84, 239–40
Sanofi Pasteur, 199
sarin gas, 135
SARS (severe acute respiratory syndrome), 159–67, 300
  control measures for, 165, 166–67
  MERS *vs.*, 168, 169
  mortality rate for, 166
  outbreaks of, 53–54, 55, 159–64
  transmission of, 164, 165–66, 167
  vaccine for, 176–77
  virus causing, 164
Sawyer, Patrick, 153–54
Schlievert, Patrick, 40
Schuchat, Anne, 166
Schweitzer, Albert, 21
Semmelweis, Ignaz, 239
serum therapy, 249
sexual assault, 75
sexually transmitted diseases, 74–75, 210–12
Shandera, Wayne, 9
Sheets, Bob, 51
Shelley, Mary, 111
Shen Jianzhong, 234–35
shingles, 65
Shin Young-soo, 250
"Shoe Leather Epidemiology," 30
Shope, Richard E., 286
Silent Spring (Carson), 193
Simonson, Stewart, 92–93, 291–92, 316
smallpox (variola major), 5
  biowarfare using, 128, 129, 130, 132–34, 135, 136–39
  eradication of, 3, 23, 24, 67
  genetic code for, 134
  host for, 61, 65
  Jenner's work on, 26–27, 80–81, 192
  Native Americans and, 66, 128
  vaccine for, 81, 82, 83, 129, 133, 136, 137, 138
smoking, campaign against, 244–45
Snow, John, 27–29, 47, 166
Song, Jae-Hoon, 173–74
Soper, Fred, 189
Sorenson, Ron, 42, 43, 45
Spellberg, Brad, 216, 222, 224–26, 241–42, 244
Staphylococcus aureus, 31, 39–40, 220
Sternberg, George Miller, 191–92
Stevens, Robert, 126
stewardship, antibiotic, 241–46
Stöhr, Klaus, 160, 161–62
Stokstad, Robert, 229
Streptococcus pneumoniae, 220, 240
streptomycin, 217
stress, adaptation to, 59, 61
sulfa drugs, 75, 216–17
Summers, Lawrence, 96, 268–69
surfactants, 37–38, 40
syphilis, 74, 75

Taleb, Nassim Nicholas, 13
Tamiflu (oseltamivir), 256, 273, 275
tampons, toxic shock syndrome and, 32, 33, 34–41
TB. *See* tuberculosis
terrorist attacks, 50, 67, 127
  domestic, 135–36
  9/11, 24, 50, 51, 125, 131, 142
  nuclear devices in, 52
  See also bioterrorism
Tesla, Nikola, 29
tetanus vaccine, 82, 83, 85
thalidomide, 209
Thompson, Tommy G., 92, 93, 126, 138, 291
threat matrix, 49–50
Timucuan Indians, 66
Todd, Jim, 30, 36
Tosh, Pritish, 173
toxic shock syndrome (TSS), 9, 30–41
trade, global, 68–70, 78, 188–89
"Tragedy of the Commons," 222–23
transmission, 72–79
  airborne, 62, 63, 73–74, 75, 76–77, 108, 117, 165
  blood-borne, 75, 101, 210
  direct-contact, 75, 77–78, 79
  food- or water-borne, 75
  global conditions and, 78–79
  reproductive rate for, 165–66
  reservoirs in, 72
  sexual, 74–75, 152, 210–12
  vector-borne, 72, 78, 79

travel, global, 68–70, 77, 146, 148, 172, 175, 211–12
Trent, William, 128
Tri-State Toxic Shock Syndrome Study (TTSSS), 33–34, 38–41
Tse, Chi Kwai, 163
tuberculosis (TB), 70, 108–10, 300
  antibiotics for, 99, 217
  drug-resistant, 108, 110, 228, 307
  global threat of, 306–8
  prevention of, 56
  statistics on, 98, 108
typhoid, 85

United Nations (UN), 105–6, 302, 307, 311
Untermeyer, Louis, 125
Urbani, Carlo, 159–60, 162

vaccines, 26, 80–84
  for *Aedes*-transmitted viruses, 202, 203
  anthrax, 94–95
  chicken pox, 83–84
  dengue, 195, 199–200
  development of, 85–97, 157–58, 176–77, 220, 248–49, 302–4
  diphtheria, 82, 83, 85
  Ebola, 89, 90, 95–96, 156–58, 177, 303
  hepatitis B, 9
  HIV, 16–17, 18–19, 89, 104, 107–8, 288, 296–97, 301
  immune response to, 16, 62–63, 294
  legacy, 196
  live virus, 83, 112–13, 122
  malaria, 89, 102
  measles, 82, 83, 288
  MERS, 95, 172, 176, 303
  movement against, 67, 84–85
  mumps, 82, 83
  pertussis, 81–82, 83, 85
  pneumococcal, 240
  polio, 82–83, 89
  rabies, 81, 84
  ring strategy for, 23, 138
  rubella, 82, 83, 210
  SARS, 176–77
  smallpox, 81, 82, 83, 129, 133, 136, 137, 138
  TB, 110
  tetanus, 82, 83, 85
  yellow fever, 85, 195–97
  Zika, 89, 90, 95, 213
  See also influenza vaccines
Vaidya, Umesh, 234
varicella-zoster virus, 65, 83–84
variola major. *See* smallpox
variolation, 80–81
vectors, 73
  control of, 26
  disease transmission by, 4, 78, 79, 182
  See also mosquitoes
Venezuelan equine encephalitis, 129
Venkayya, Rajeev, 89
Venter, J. Craig, 134
viruses, 61–62
  bacteriophage, 249
  engineered, 112–13, 115–16, 117
  live attenuated, 83, 122
  reproduction by, 62
  transmission of, 73, 76–79, 151–52

Wagenaar, Jaap, 231
Wakefield, Andrew, 288
Waksman, Selman, 217
Wallace, Mike, 125
Walsh, Timothy, 235
war
  biological weapons in, 114, 127–28
  disease compared to, 5–6, 53, 91–92, 269, 283, 317
  rape as weapon of, 75
  thermonuclear, 52
  See also bioterrorism
Washington, George, 81
Wasserman, Sean, 195
water
  contaminated, 27–28, 75, 240
  supply of, 239–40, 300, 308–9
Watson, James, 133
weapons of mass destruction, 130–31, 135
Webster, Robert, 260
Weems, Kerry, 93
Wellcome Trust, 158, 218, 313
Wellstone, Paul, 126

Western equine encephalitis, 182, 186–88
West Nile virus, 78, 182
White, T. H., 292
Whitehead, Henry, 27–28
Wimmer, Eckard, 133, 134
Wolbachia, 203
Wood, Leonard, 192
World Bank Pandemic Emergency Financing Facility, 53
World Economic Forum, 97, 218, 252–53
World Health Assembly, 240, 311, 312
World Health Organization (WHO)
  Aedes-transmitted diseases and, 204, 305
  dengue and, 199
  Ebola and, 145, 310, 312
  Global Health Security Agenda and, 315
  global support for, 310
  HIV guidelines of, 105
  influenza and, 264–65, 283–84, 301
  MERS road map of, 176
  One Health approach of, 309
  overhaul of, 311–14
  SARS and, 160, 161–64, 176
  vaccines and, 95, 96–97, 240, 303
  yellow fever and, 194, 195, 197
World Organisation for Animal Health (OIE), 265, 309
yellow fever, 190–97, 300
  eradication of, 189
  global response to, 315
  mosquito control for, 192–93
  outbreaks of, 54, 78, 194–95, 312
  sexual transmission of, 210
  vaccine for, 85, 195–97
  vector for, 101, 188, 189, 191–92, 200

Zaki, Ali Mohamed, 167
zanamivir (Relenza), 256, 273, 275, 276
Zika, 4, 205–14, 300
  complications of, 205, 206–10, 211
  diagnostic tests for, 304
  funding for, 107, 157, 212, 317, 318
  genetic changes in, 212–13
  global response to, 315
  outbreaks of, 47–48, 54, 157, 205, 207
  sexual transmission of, 152, 210–12
  vaccine for, 89, 90, 95, 213
  vector for, 78, 79, 101, 189, 200, 205, 207, 209, 213
zoonotic diseases, 3–4, 64–65
  See also animal reservoirs
Zosyn (piperacillin-tazobactam), 224–25, 244

About the Authors

MICHAEL T. OSTERHOLM, PhD, MPH, is Regents Professor, McKnight Presidential Endowed Chair in Public Health, and the founding director of the Center for Infectious Disease Research and Policy (CIDRAP) at the University of Minnesota. An internationally renowned epidemiologist and former state epidemiologist of Minnesota and chief of the Acute Disease Epidemiology Section, he has led many investigations of outbreaks of international importance, including foodborne diseases, the association of tampons and toxic shock syndrome, the transmission of hepatitis B in healthcare settings, and HIV infection in healthcare workers.

Dr. Osterholm is the author, with John Schwartz, of the *New York Times* best-selling *Living Terrors: What America Needs to Know to Survive the Coming Bioterrorist Catastrophe,* as well as more than 315 papers and abstracts, including twenty-one book chapters. He serves on the editorial boards of nine journals, has been an international leader regarding preparedness for an influenza pandemic, and has sounded the alarm regarding critical infectious disease threats in *Foreign Affairs,* the *New England Journal of Medicine,* and *Nature* as well as the op-ed pages of the *New York Times* and the *Washington Post.* He has been an international leader on the growing concern regarding the use of biological agents as catastrophic weapons targeting civilian populations. In that role, he served as a personal adviser to the late king Hussein of Jordan.

He served as a special adviser to HHS secretary Tommy Thompson. Thompson's successor, Secretary Michael Leavitt, appointed Osterholm to the newly established National Science Advisory Board on Biosecurity. He is a member on the National Academy of Medicine, the Council on Foreign Relations, and numerous other professional organizations. He and his wife, Libby, live in Minneapolis.

MARK OLSHAKER is an Emmy Award–winning documentary filmmaker and a *New York Times* number one best-selling author of five novels and ten books of nonfiction. His books with former FBI special agent and profiling pioneer John Douglas, beginning with *Mindhunter*—now an original Netflix series—and up through the recent *Law & Disorder,* have sold millions of copies, been translated into many languages, and offer a unique and intriguing perspective into behavioral science and criminal investigative analysis.

His scientific and medical writing is represented by *Virus Hunter: Thirty Years of Battling Hot Viruses Around the World,* with Dr. C. J. Peters, which the *New York Times* placed on its Noteworthy and Recommended lists for the year and which the *New England Journal of Medicine* compared to Paul de Kruif's celebrated *Microbe Hunters,* declaring it "not merely the exhilarating tale of three decades of scientific research. It is also an outspoken, comprehensive analysis of the political and human issues that front-line scientists fighting outbreaks of hemorrhagic fever deal with daily." Olshaker wrote *The Instant Image: Edwin Land and the Polaroid Experience,* the lead chapter of the medical textbook *Forensic Emergency Medicine,* the IMAX film *Stormchasers,* and the PBS programs *What's Killing the Children?, Bioterror: Living with a New Reality,* and *Anatomy of a Pandemic,* among numerous others.

Olshaker's highly praised suspense novels include *Einstein's Brain, Unnatural Causes, Blood Race,* and *The Edge,* which *Publish-*

ers Weekly called "a darkly imagined thriller marked by brisk action and a mind-bending denouement."

He is the former chairman of the Cosmos Club Foundation and serves on the boards of the Norman Mailer Society and the Rod Serling Memorial Foundation. Olshaker and his wife, Carolyn, an attorney, live in Washington, DC.